Android项目实战
——手机安全卫士 精彩模块欣赏

第1章 欢迎界面和主界面

第4章 软件管家模块

第2章 手机防盗

第2章 手机防盗模块

第3章 通讯卫士模块

第5章 手机杀毒模块

第6章 缓存清理模块

第9章 高级工具模块

Android项目实战——手机安全卫士 精彩模块欣赏

第7章 进程管理模块

第8章 流量统计模块

第10章 设置中心模块

第9章 高级工具模块

NITE 国家信息技术紧缺人才培养工程指定教材

Android 项目实战——手机安全卫士

传智播客高教产品研发部　编著

中国铁道出版社有限公司
CHINA RAILWAY PUBLISHING HOUSE CO., LTD.

内 容 简 介

本书以项目为导向,通过"手机安全卫士"项目讲解了一个完整的 Android 项目开发流程。全书共 10 章,每章针对一个功能模块进行详细讲解,例如第 2 章讲解的是手机防盗模块,其功能包含 SIM 卡变更提醒、GPS 定位、远程锁屏等。第 3 章讲解的是通讯卫士模块,其功能包含添加黑名单、删除黑名单、短信拦截、电话拦截等。通过本书内容的学习,可以让更多编程者快速积累开发经验,具备中级 Android 工程师的能力。

本书附有配套的教学课件、源代码、习题、面试题、教学案例等资源,而且为了帮助编程者更好地学习书中内容,还提供了在线答疑服务,希望得到更多读者的关注。

本书可作为高等院校本、专科计算机相关专业程序设计类课程的专用教材,也可作为 Android 爱好者的自学教材。

图书在版编目(CIP)数据

Android 项目实战:手机安全卫士 / 传智播客高教产品研发部编著. — 北京:中国铁道出版社,2015.8(2019.7 重印)

ISBN 978-7-113-20549-2

Ⅰ. ①A… Ⅱ. ①传… Ⅲ. ①移动终端-应用程序-程序设计 Ⅳ. ①TN929.53

中国版本图书馆 CIP 数据核字(2015)第 144976 号

书 名:	Android 项目实战——手机安全卫士
作 者:	传智播客高教产品研发部 编著

策 划:	翟玉峰	读者热线:(010)63550836
责任编辑:	翟玉峰 王 惠	
封面设计:	徐文海	
封面制作:	白 雪	
责任校对:	汤淑梅	
责任印制:	郭向伟	

出版发行:中国铁道出版社有限公司(100054,北京市西城区右安门西街 8 号)
网 址:http://www.tdpress.com/51eds/

印 刷:三河市宏盛印务有限公司
版 次:2015 年 8 月第 1 版 2019 年 7 月第 6 次印刷
开 本:787mm×1092mm 1/16 印张:20.25 彩插:2 字数:490 千
印 数:17 001~19 000 册
书 号:ISBN 978-7-113-20549-2
定 价:39.80 元

版权所有 侵权必究

凡购买铁道版图书,如有印制质量问题,请与本社教材图书营销部联系调换。电话:(010)63550836

打击盗版举报电话:(010)51873659

PREFACE 序

江苏传智播客教育科技股份有限公司（简称传智播客）是一家致力于培养高素质软件开发人才的科技公司，"黑马程序员"是传智播客旗下高端IT教育品牌。

"黑马程序员"的学员多为大学毕业后，想从事IT行业，但各方面条件还不成熟的年轻人。"黑马程序员"的学员筛选制度非常严格，包括了严格的技术测试、自学能力测试，还包括性格测试、压力测试、品德测试等。百里挑一的残酷筛选制度确保学员质量，并降低企业的用人风险。

自"黑马程序员"成立以来，教学研发团队一直致力于打造精品课程资源，不断在产、学、研三个层面创新自己的执教理念与教学方针，并集中"黑马程序员"的优势力量，有针对性地出版了计算机系列教材80多种，制作教学视频数十套，发表各类技术文章数百篇。

"黑马程序员"不仅斥资研发IT系列教材，还为高校师生提供以下配套学习资源与服务。

为大学生提供的配套服务

（1）请登录在线平台：http://yx.ityxb.com，进入"高校学习平台"，免费获取海量学习资源，帮助高校学生解决学习问题。

（2）针对高校学生在学习过程中存在的压力等问题，我们还面向大学生量身打造了IT技术女神——"播妞学姐"，可提供教材配套源码和习题答案以及更多IT其他干货资源，同学们快来关注"播妞学姐"微信公众号：boniu1024。

"播妞学姐"微信公众号

为教师提供的配套服务

针对高校教学，"黑马程序员"为IT系列教材精心设计了"教案+授课资源+考试系统+题库+教学辅助案例"的系列教学资源，高校老师请登录在线平台：http://yx.ityxb.com 进入"高校教辅平台"或关注码大牛老师微信/QQ：2011168841，获取配套资源，也可以扫描下方二维码，加入专为IT教师打造的师资服务平台——"教学好助手"，获取最新教师教学辅助资源的相关动态。

前言

为什么要学习 Android

Android 是 Google 公司开发的基于 Linux 的开源操作系统,主要应用于智能手机、平板电脑等移动设备。经过短短几年的发展,Android 系统在全球得到了大规模推广,除智能手机和平板电脑外,还可用于穿戴设备、智能家具等领域。据不完全统计,Android 系统已经占据了全球智能手机操作系统的 80%以上,中国市场占有率更是高达 90%以上。由于 Android 的迅速发展,导致市场对 Android 开发人才需求猛增,因此越来越多的人学习 Android 技术,以适应市场需求,寻求更广阔的发展空间。

如何使用本书

本书以项目为导向,通过手机安全卫士讲解了一个完整的项目开发流程,该项目不仅涵盖了市面上所有主流手机卫士的功能,同时也是对 Android 基础知识的一个综合运用,因此,本书更适合具备一定 Android 基础并需要提高项目经验的开发人员使用。

本教材共 10 章,每个章节针对一个功能模块进行讲解,接下来分别进行简要介绍,具体如下:

● 第 1 章项目简介,主要针对手机安全卫士进行项目分析、功能展示以及实现欢迎界面和主界面开发。

● 第 2 章手机防盗模块,主要针对手机 SIM 卡绑定、GPS 定位、远程锁屏、远程删除数据等进行讲解。

● 第 3 章通讯卫士模块,主要针对黑名单添加、电话拦截、短信拦截等进行讲解。

● 第 4 章软件管家模块,主要针对软件的快速启动、卸载、分享、设置等进行讲解。

● 第 5 章手机杀毒模块,主要针对第 3 方数据库的使用、病毒扫描、病毒查杀等进行讲解。

● 第 6 章缓存清理模块,主要针对如何获取手机中的缓存信息,并对缓存清理等进行讲解。

● 第 7 章进程管理模块,主要针对如何获取手机中正在运行的进程、结束进程进行讲解。

● 第 8 章流量统计模块,主要针对运营商信息设置、获取流量套餐、显示本月和本日流量等进行讲解。

● 第 9 章高级工具模块,主要针对号码归属地查询、短信备份、短信还原、程序锁等进行讲解。

● 第 10 章设置中心模块,主要针对手机卫士设置进行讲解,如是否开启黑名单拦截功能、是否开启程序锁功能等。

致谢

本教材的编写和整理工作由传智播客教育科技有限公司高教产品研发部完成,主要参与人员有徐文海、陈欢、阳丹、安鹏宇、张鑫、李健、郝丽新、柴永菲、张泽华、李印东、刘亚超、邱本超、殷凯、马伟奇、刘峰、金兴,全体人员在这近一年的编写过程中付出了很多辛勤的汗水。除此之外,还有传智播客 600 多名学员也参与到了教材的试读工作中,他们站在初学者的角度对教材提供了许多宝贵的修改意见,在此一并表示衷心的感谢。

意见反馈

尽管我们尽了最大的努力,但教材中难免会有不妥之处,欢迎各界专家和读者朋友们来信来函给予宝贵意见,我们将不胜感激。您在阅读本书时,如发现任何问题或有不认同之处可以通过电子邮件与我们取得联系。

请发送电子邮件至:itcast_book@vip.sina.com

传智播客教育科技有限公司高教产品研发部
2015-6-1 于北京

目　录

第 1 章　项目简介 ……………………1
1.1　项目概述 …………………………1
1.1.1　项目分析 ……………………1
1.1.2　功能展示 ……………………2
1.1.3　代码结构 ……………………7
1.2　欢迎界面 …………………………7
1.2.1　开发流程图 …………………8
1.2.2　欢迎界面 UI …………………8
1.2.3　服务器的搭建 ………………9
1.2.4　下载和安装 APK ……………11
1.2.5　版本更新工具类 ……………15
1.2.6　版本信息的实体类 …………20
1.2.7　欢迎界面逻辑 ………………20
1.3　主界面 ……………………………21
1.3.1　主界面 UI ……………………21
1.3.2　主界面 Item 布局 ……………23
1.3.3　数据适配器 …………………23
1.3.4　主界面逻辑 …………………25
本章小结 ………………………………27

第 2 章　手机防盗模块 ………………28
2.1　模块概述 …………………………28
2.1.1　功能介绍 ……………………28
2.1.2　开发流程图 …………………30
2.1.3　代码结构 ……………………30
2.2　设置密码 …………………………31
2.2.1　设置密码界面 ………………31
2.2.2　自定义对话框样式 …………34
2.2.3　设置密码逻辑 ………………34
2.2.4　MD5 加密算法 ………………36
2.3　输入密码 …………………………37
2.3.1　输入密码界面 ………………37
2.3.2　输入密码逻辑 ………………39
2.4　设置向导界面 ……………………40
2.4.1　小圆点界面 …………………40

2.4.2　向导界面（一） 42
　　　2.4.3　向导界面（二） 46
　　　2.4.4　向导界面（三） 49
　　　2.4.5　向导界面（四） 51
　　　2.4.6　指令界面 53
　2.5　设置向导功能 59
　　　2.5.1　滑屏动画 59
　　　2.5.2　手势滑动 61
　　　2.5.3　向导功能（一） 63
　　　2.5.4　向导功能（二） 64
　　　2.5.5　向导功能（三） 68
　　　2.5.6　获取联系人 70
　　　2.5.7　向导功能（四） 78
　　　2.5.8　防盗指令 80
　　　2.5.9　修改 HomeActivity 文件 87
　本章小结 93

第3章　通讯卫士模块 94

　3.1　模块概述 94
　　　3.1.1　功能介绍 94
　　　3.1.2　开发流程图 95
　　　3.1.3　代码结构 96
　3.2　黑名单数据库 97
　　　3.2.1　创建数据库 97
　　　3.2.2　联系人的实体类 98
　　　3.2.3　数据库操作类 98
　　　3.2.4　测试数据 102
　3.3　主界面 105
　　　3.3.1　主界面 UI 105
　　　3.3.2　黑名单 Item 布局 107
　　　3.3.3　主界面逻辑代码 109
　　　3.3.4　数据适配器 112
　3.4　添加黑名单 115
　　　3.4.1　添加黑名单界面 115
　　　3.4.2　添加黑名单逻辑 118
　　　3.4.3　联系人列表 121
　3.5　黑名单拦截 122
　　　3.5.1　拦截短信 122
　　　3.5.2　拦截电话 123
　本章小结 127

第4章　软件管家模块 128

　4.1　模块概述 128
　　　4.1.1　功能介绍 128
　　　4.1.2　代码结构 128
　4.2　软件管家界面 129
　　　4.2.1　软件管家 UI 129
　　　4.2.2　软件管家 Item 布局 131
　　　4.2.3　应用程序的实体类 134
　4.3　工具类 134
　　　4.3.1　获取应用程序信息 134
　　　4.3.2　单位转换 136
　　　4.3.3　程序的业务类 137
　4.4　软件管家功能 139
　　　4.4.1　软件管家逻辑 139
　　　4.4.2　数据适配器 144
　本章小结 150

第5章　手机杀毒模块 151

　5.1　模块概述 151
　　　5.1.1　功能介绍 151
　　　5.1.2　代码结构 152
　　　5.1.3　手机病毒 152
　5.2　数据库操作 153
　　　5.2.1　数据库展示 153
　　　5.2.2　数据库操作 154
　　　5.2.3　获取 MD5 码 154
　5.3　病毒查杀 156
　　　5.3.1　病毒查杀界面 156
　　　5.3.2　病毒查杀逻辑代码 158
　5.4　查杀进度 160
　　　5.4.1　查杀进度界面 160
　　　5.4.2　查杀进度 Item 布局 162
　　　5.4.3　病毒的实体类 163
　　　5.4.4　查杀进度逻辑 164
　　　5.4.5　数据适配器 169
　本章小结 171

第6章　缓存清理模块 172

　6.1　模块概述 172
　　　6.1.1　功能介绍 172

		6.1.2	代码结构 172
	6.2	扫描缓存 173
		6.2.1	扫描缓存界面 173
		6.2.2	缓存清理 Item 布局 175
		6.2.3	缓存信息的实体类 176
		6.2.4	扫描缓存逻辑 177
		6.2.5	数据适配器 181
	6.3	缓存清理 183
		6.3.1	缓存清理界面 183
		6.3.2	缓存清理逻辑 186
	本章小结 191

第 7 章 进程管理模块 192

	7.1	模块概述 192
		7.1.1	功能介绍 192
		7.1.2	代码结构 192
	7.2	进程管理 193
		7.2.1	进程管理界面 193
		7.2.2	进程管理 Item 布局 196
		7.2.3	进程信息的实体类 197
		7.2.4	主界面逻辑 198
		7.2.5	数据适配器 204
	7.3	工具类 208
		7.3.1	获取系统信息 208
		7.3.2	获取进程信息 210
	7.4	设置进程 211
		7.4.1	设置进程界面 212
		7.4.2	设置进程逻辑 214
		7.4.3	锁屏清理进程服务 216
	本章小结 217

第 8 章 流量统计模块 218

	8.1	模块概述 218
		8.1.1	功能介绍 218
		8.1.2	代码结构 219
	8.2	运营商设置 219
		8.2.1	运营商设置界面 219
		8.2.2	运营商设置逻辑 222
	8.3	数据库操作 223
		8.3.1	创建数据库 224

		8.3.2	数据库操作 224
	8.4	流量监控 226
		8.4.1	流量监控界面 226
		8.4.2	流量监控逻辑 229
		8.4.3	判断服务是否运行 234
		8.4.4	获取流量的服务 235
		8.4.5	开机广播 238
	本章小结 238

第 9 章 高级工具模块 239

	9.1	模块概述 239
		9.1.1	功能介绍 239
		9.1.2	代码结构 241
	9.2	主界面 242
		9.2.1	自定义组合控件 242
		9.2.2	主界面逻辑 247
	9.3	号码归属地查询 249
		9.3.1	号码归属地查询界面 249
		9.3.2	数据库展示 250
		9.3.3	数据库操作 251
		9.3.4	号码归属地查询逻辑 254
	9.4	短信备份 257
		9.4.1	短信备份工具类 257
		9.4.2	短信加密和解密 261
		9.4.3	短信备份界面 262
		9.4.4	短信备份逻辑 267
		9.4.5	Toast 封装 270
	9.5	短信还原 271
		9.5.1	短信还原工具类 271
		9.5.2	短信的实体类 274
		9.5.3	短信还原界面 274
		9.5.4	短信还原逻辑 275
	9.6	程序锁 277
		9.6.1	创建数据库 277
		9.6.2	数据库操作类 278
		9.6.3	获取所有应用工具类 280
		9.6.4	应用的实体类 281
		9.6.5	程序锁界面 281
		9.6.6	程序锁逻辑 283

9.6.7　加锁与未加锁功能........286
　　　9.6.8　程序锁服务..................295
　9.7　密码锁..299
　　　9.7.1　密码锁界面..................299
　　　9.7.2　密码锁逻辑..................300
　本章小结...303
第10章　设置中心模块........................304
　10.1　模块概述.....................................304

　　　10.1.1　功能介绍......................304
　　　10.1.2　代码结构......................304
　10.2　设置中心.....................................305
　　　10.2.1　自定义控件..................305
　　　10.2.2　设置中心界面..............310
　　　10.2.3　工具类..........................311
　　　10.2.4　设置中心逻辑..............312
　本章小结...314

➡ 项目简介

学习目标
- 了解手机卫士的项目结构;
- 掌握欢迎界面以及程序主界面的开发。

在智能手机的世界里还能保护自己的隐私吗？平日里小心谨慎，可是隐私信息被盗还是频有发生。为了防止隐私泄露，我们开发了一个手机安全卫士软件，该软件主要用于管理手机中的数据以及保护用户隐私。本章将针对手机安全卫士中的欢迎界面以及程序主界面进行详细讲解。

1.1 项目概述

1.1.1 项目分析

手机安全卫士主要分为 9 个功能模块，其中包含手机防盗、通讯卫士、软件管家、手机杀毒、缓存清理、进程管理、流量统计、高级工具、设置中心，如图 1-1 所示。

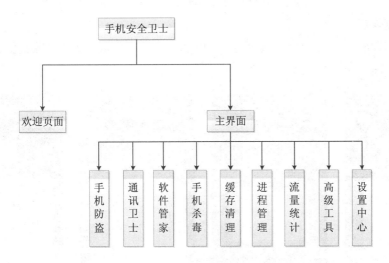

图 1-1 项目结构

从图 1-1 可以看出，手机安全卫士主要分为两块：一个是欢迎界面；另一个是主界面。在欢迎界面中会显示程序的版本号以及功能提示等，在主界面中显示 9 个功能模块，每个功能模块还有具体的小功能。

1.1.2 功能展示

Android 手机卫士的功能比较多，为了方便大家熟悉每个功能的作用，接下来针对各模块进行展示，具体如下：

1. 欢迎界面和主界面

程序的欢迎界面即 Splash 界面，会显示程序的 Logo 以及版本信息等，如果服务器的版本号与本地版本号一致，则直接进入主界面，否则会弹出提示信息询问是否升级。如果点击"立即升级"按钮，会从服务端下载最新版本的手机安全卫士 APP，下载完成后直接进行安装，如果点击"暂不升级"按钮则直接进入主界面。欢迎界面和主界面如图 1-2 所示。

图 1-2　欢迎界面和主界面

2. 手机防盗模块

手机防盗模块主要用于 SIM 变更提醒、GPS 追踪、远程锁屏、远程删除数据等。该模块界面效果如图 1-3 所示。

图 1-3　手机防盗界面

图 1-3　手机防盗界面（续）

3. 通讯卫士模块

通讯卫士模块用于实现黑名单拦截功能，对黑名单中的号码进行短信或电话拦截。该模块界面效果如图 1-4 所示。

图 1-4　通讯卫士界面

4. 软件管家模块

软件管家模块主要用于管理软件的启动、卸载、分享、设置等。该模块界面效果如图1-5所示。

图1-5 软件管家界面

5. 手机杀毒模块

手机杀毒模块主要用于全盘扫描，并显示当前正在扫描的病毒以及查杀进度。该模块界面效果如图1-6所示。

图1-6 手机杀毒界面

6. 缓存清理模块

缓存清理模块主要用于查看所有程序的缓存，并可以一键清理所有程序的缓存，该模块界面效果如图1-7所示。

7. 进程管理模块

进程管理模块主要用于查看手机中正在运行的进程信息，以及选中清理进程等。该模块界面效果如图1-8所示。

图 1-7　缓存清理界面

图 1-8　进程管理界面

8. 流量统计模块

流量统计模块主要用于显示运营商信息设置、流量监控，在流量监控界面中可以看见本日、本月使用流量以及本月的总流量。该模块界面效果如图 1-9 所示。

图 1-9　流量监控界面

9. 高级工具模块

高级工具模块主要包括号码归属地查询、短信备份、短信还原和程序锁四个功能。该模块界面效果如图 1-10 所示。

图 1-10 高级工具界面

10. 设置中心模块

设置中心模块主要用于设置黑名单拦截是否开启、程序锁是否开启。该模块界面效果如图 1-11 所示。

图 1-11 设置中心界面

以上对手机安全卫士项目的所有功能进行了效果展示，接下来就进入项目的正式开发阶段。在学习该项目时，编程者一定要动手完成每一个功能模块，熟练掌握项目的核心代码。

1.1.3 代码结构

在开发项目时，均会按照功能将其分类放在不同的包中，本教材以章节为编号作为包名将每个大的功能模块放在一个包中，例如第 1 章代码全部放在 chapter01 包中，在这个包中还可以根据功能划分出 adapter 包、entity 包、utils 包等。通常情况下，操作界面的 Activity 都直接放在章节包中（如 chapter01），主界面的 HomeActivity 文件在开发每个功能模块时都会使用，因此将其直接放在 cn.itcast.mobliesafe 包中。

为了让大家更清楚该模块结构，接下来下面给出图例来展示欢迎界面与主界面的代码结构，如图 1-12 所示。

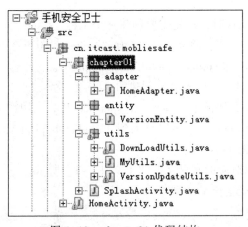

图 1-12　chapter01 代码结构

下面按照结构顺序依次介绍 chapter01 包中的文件，具体如下：
- HomeAdapter.java：主界面的布局填充器，用于填充界面中的 9 个功能图标以及文字信息；
- VersionEntity.java：封装版本信息的实体类，主要包含字段 versioncode、description、apkurl；
- DownLoadUtils.java：用于从服务器下载 APK 的工具类；
- MyUtils.java：用于获取应用程序的版本信息，并实现新版本 APK 的安装；
- VersionUpdateUtils.java：用于进行版本更新的工具类；
- SplashActivity.java：用于展示欢迎界面以及版本信息；
- HomeActivity.java：用于展示程序主界面以及实现各个功能图标的点击事件。

1.2　欢迎界面

在每个应用程序中欢迎界面都是必不可少的，它的主要作用是展示产品 Logo、检查程序完整性、检查程序的版本更新、加载广告页、做一些初始化操作等。本节将针对欢迎界面开发进行详细讲解。

1.2.1 开发流程图

在程序开发中，使用流程图可以更好地分析程序开发流程。接下来展示一下欢迎界面的开发流程图，具体如图 1-13 所示。

从图 1-13 可以看出，欢迎页面需要获取应用程序的本地版本号，并与服务器中应用程序版本比对，若版本号相同则进入主界面，若不相同则弹出版本升级对话框，让用户选择是否更新应用程序版本，如果选择更新则下载安装新版本 APK 文件，否则立即进入程序主界面。

1.2.2 欢迎界面 UI

在开发手机安全卫士项目时，首先需要创建一个工程，将其命名为"手机安全卫士"，将包名指定为"cn.itcast.mobliesafe"，然后创建第一个包 cn.itcast.mobliesafe.chatper01。由于欢迎界面就是 Splash 界面，因此将要创建的 Activity 命名为"SplashActivity"，布局文件指定为"activity_splash.xml"，然后将欢迎界面的背景图（launch_bg.jpg）导入到 drawable-hdpi 目录中。欢迎界面的图形化界面如图 1-14 所示。

图 1-14 所示欢迎界面对应的布局文件如【文件 1-1】所示。

图 1-13 欢迎界面开发流程图

图 1-14 欢迎界面

【文件 1-1】activity_splash.xml

```
<RelativeLayout xmlns:android="http://schemas.android.com/apk/res/android"
    xmlns:tools="http://schemas.android.com/tools"
    android:layout_width="match_parent"
    android:layout_height="match_parent"
    android:background="@drawable/launch_bg"
    tools:context=".SplashActivity" >
<ProgressBar
    android:id="@+id/progressBar1"
    android:layout_width="wrap_content"
    android:layout_height="wrap_content"
    android:layout_centerVertical="true"
    android:layout_centerHorizontal="true" />
<TextView
    android:id="@+id/tv_splash_version"
    android:layout_width="wrap_content"
    android:layout_height="wrap_content"
    android:layout_centerHorizontal="true"
```

```xml
        android:layout_below="@+id/progressBar1"
        android:layout_marginTop="10dp"
        android:textColor="#A2A2A2"
        android:shadowColor="#ffff00"
        android:shadowDx="1"
        android:shadowDy="1"
        android:shadowRadius="2"
        android:text="版本号:1.0"
        android:textSize="16sp" />
</RelativeLayout>
```

上述布局文件中，使用的是 RelativeLayout 布局，该布局中放置了一个 ProgressBar 控件和一个 TextView 控件，ProgressBar 控件用于显示程序加载的进度条，TextView 控件用于显示程序的版本号。

需要注意的是，TextView 中有几个特殊属性用于指定文字阴影效果，其中 android:shadowColor 属性用于指定阴影颜色，android:shadowDx、android:shadowDy、android:shadowRadius 属性分别用于指定阴影在 X 轴和 Y 轴上的偏移量以及阴影的半径。阴影的半径必须设置，当数值为 0 时无阴影，数值越大时阴影会越透明，扩散效果越明显。

1.2.3 服务器的搭建

由于本程序需要获取服务端应用的版本号以及下载服务器最新的 APK，因此需要搭建一个服务器。搭建步骤如下：

（1）本程序使用 Tomcat 作为服务器，点击 Tomcat 目录下的 bin/startup.bat 开启服务器。

（2）创建一个 HTML 页面（updateinfo.html），该页面返回的信息需要包括服务器中 APK 的版本号、版本说明以及新版本的下载地址。updateinfo.html 页面的 JSON 信息如下所示：

```
{
    "code":"2.0",
    "des":"手机卫士2.0版本，新增了手机杀毒功能",
    "apkurl":"http://172.16.25.13:8080/mobliesafe.apk"
}
```

（3）将 updateinfo.html 页面以及手机安全卫士 2.0 的 APK（经过签名打包用于发布的 APK，而不是调试的 APK）复制到 Tomcat 的 webapps/ROOT 文件夹下。

> **多学一招：JSON 数据**
>
> JSON 即 JavaScript Object Notation（对象表示法），是一种轻量级的数据交换格式，它基于 JavaScript 的一个子集，使用了类似于 C 语言家族的习惯（包括 C、C++、C#、Java、JavaScript、Perl、Python 等）。这些特性使 JSON 成为理想的数据交互语言，易于阅读和编写，同时也易于机器解析和生成。
>
> 与 XML 一样，JSON 也是基于纯文本的数据格式，并且 JSON 的数据格式非常简单，既可以用 JSON 传输一个简单的 String、Number、Boolean 类型数据，也可以传输一个数组，或者一个复杂的 Object 对象。

JSON 有两种结构，具体如下：

- 值的有序列表：在大部分语言中，它被理解为数组（array）。
- "名称/值"对的集合：在不同的语言中，它被理解为对象、记录、结构、字典、哈希表等。

这些都是常见的数据结构，事实上大部分现代计算机语言都以某种形式支持它们。这使得一种数据格式在不同的编程语言之间交互成为可能。

使用 JSON 表示数组时，数组以"["开始，以"]"结束，每个元素之间使用","（逗号）分隔（元素可以是任意的 Value）。其存储形式如图 1-15 所示。

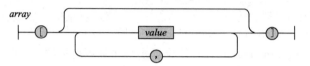

图 1-15　存储数组

例如，一个数组包含了 String、Number、Boolean、null 类型数据，JSON 的表示形式如下：

```
["abc",12345,false,null]
```

使用 JSON 表示 Object 类型数据时，Object 对象以"{"开始，以"}"结束，每个"名称"后跟一个":"（冒号）；"名称/值"对之间使用","（逗号）分隔。其存储形式如图 1-16 所示。

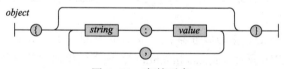

图 1-16　存储对象

例如，一个 address 对象包含城市、街道、邮编等信息，JSON 的表示形式如下：

```
{"city":"Beijing","street":"Chaoyang Road","postcode":100025 }
```

当使用 JSON 存储 Object 时，其中的 Value 既可以是一个 Object，也可以是数组，因此，复杂的 Object 可以嵌套表示。例如，一个 Person 对象包含 name 和 address 对象，其表示格式如下：

```
{
    "name":"Michael",
    "address":{
        "city":"Beijing",
        "street":"Chaoyang Road",
        "postcode":100025
    }
}
```

假设 Value 值是一个数组，例如，一个 Person 对象包含 name 和 hobby 信息，其表示格式如下：

```
{
    "name":"Michael",
    "hobby":["篮球","羽毛球","游泳"]
}
```

需要注意的是，如果使用 JSON 存储单个数据（如"abc"），一定要使用数组的形式，不要使用 Object 形式，因为 Object 形式必须是"名称/值"的形式。

1.2.4 下载和安装 APK

1. 下载 APK

本项目采用第三方开源框架 xUtils 下载 APK，因此需要将 xUtils 的 jar 包复制到工程 libs 目录下，选中 xUtils 工具包并右击，选择 Build Path→Add to Build Path 命令将 jar 包导入工程。

xUtils 是 Android 的第三方开源框架，它起源于 Afinal 框架（用于发送 HTTP 请求、显示 Bitmap 图片等），同时 xUtils 包含了很多实用的 Android 工具类。接下来介绍一下本项目用到 xUtils 的几个类和方法，具体如下所示：

（1）HttpUtils 类用于发送 HTTP 请求、上传文件、下载文件等，其中 HttpUtils 的 download(String url,String target,RequestCallBack<File> requestCallBack)方法是用来下载文件的。参数 url 代表要下载文件的路径，target 代表下载文件的本地路径，requestCallBack 是一个接口对象用于监听文件下载状态的。

（2）RequestCallBack<File>接口有三个抽象方法，分别在下载成功、下载失败、下载中调用，通过这三个抽象方法可以获取到文件的下载状态。

在 cn.itcast.mobliesafe.chapter01 中创建 utils 包，用于存放下载文件的工具类 DownLoadUtils.java，具体代码如【文件 1-2】所示。

【文件 1-2】DownLoadUtils.java

```
1  public class DownLoadUtils{
2      /**
3       * 下载 APK 的方法
4       * @param url
5       * @param targerFile
6       * @param myCallBack
7       */
8      public void downapk(String url,String targerFile,final MyCallBack myCallBack){
9          //创建 HttpUtils 对象
10         HttpUtils httpUtils=new HttpUtils();
11         //调用 HttpUtils 下载的方法下载指定文件
12         httpUtils.download(url,targerFile,new RequestCallBack<File>(){
13             @Override
14             public void onSuccess(ResponseInfo<File> arg0){
```

```
15              myCallBack.onSuccess(arg0);
16          }
17          @Override
18          public void onFailure(HttpException arg0,String arg1){
19              myCallBack.onFailure(arg0,arg1);
20          }
21          @Override
22          public void onLoading(long total,long current,boolean isUploading){
23              super.onLoading(total,current,isUploading);
24              myCallBack.onLoadding(total,current,isUploading);
25          }
26      });
27  }
28  /**
29   *接口，用于监听下载状态的接口
30   */
31  interface MyCallBack{
32      /**下载成功时调用*/
33      void onSuccess(ResponseInfo<File> arg0);
34      /**下载失败时调用*/
35      void onFailure(HttpException arg0,String arg1);
36      /**下载中调用*/
37      void onLoadding(long total,long current,boolean isUploading);
38  }
39 }
```

代码说明：

- 第 8~27 行的 downapk()方法用于下载 APK，在该方法中创建 HttpUtils 对象，然后调用它指定的下载方法 download()，在调用该方法时，需要实现 RequestCallBack<File> 接口中的三个抽象方法。
- 第 31~38 行的 MyCallBack 回调接口用于监听文件的下载状态，它的作用与 Resquest CallBack 的作用一致。

多学一招：回调函数

在学习 Android 过程中，经常会遇到"回调函数"这个词，那么什么是回调函数呢？简单地说，回调函数就是通过其指针来调用的函数，它不会被自己所在的对象调用，只会在调用别人的方法的时候反过来被调用。大家都知道，Android 程序是通过 Java 语言来实现的，Java 中是没有指针的，因此在实现回调时都是通过接口或抽象类。

回调的过程简单理解为，在 A 类中定义了一个方法，这个方法中用到了一个接口和该接口中的抽象方法，但是抽象方法没有具体的实现，需要 B 类去实现。当 B 类实现该方法后，它本身不会去调用该方法，而是传递给 A 类，供 A 类去调用。这种机制就称为回调。

回调机制是将实现功能和定义分离的一种手段，是一种松耦合的设计思想。接下来通过一段代码进行分析，具体如下：

(1) 定义回调接口 ICallBack

```java
//声明一个接口
public interface ICallBack{
  void postExec();
}
```

(2) 定义一个实现类 FooBar

```java
public class FooBar{
  private ICallBack callBack;
  public void setCallBack(ICallBack callBack){
     this.callBack=callBack;
  }
  public void doSth(){
     callBack.postExec();
  }
   public static void main(String[] args){
      FooBar foo=new FooBar();
      foo.setCallBack(new ICallBack(){
         public void postExec(){
             Log.i("info","method executed.");
         }
      });
      foo.doSth();//调用函数
   }
}
```

在上述代码中，第一段代码定义了一个回调接口 ICallBack，该接口中有一个 postExec() 方法，但是并没有实现。第二段代码定义了一个 FooBar 类，该类中有一个 setCallBack() 方法，接收一个 ICallBack 参数，然后在 doSth() 方法中通过 callBack.postExec() 实现 postExec() 方法，最后在 main() 方法中进行调用。这就是一个回调函数的基本用法。

在 Android 开发中，回调函数使用非常广泛，下面列举两个回调函数的使用场景，让大家更直观地看到回调函数是如何应用的。

应用场景一：事件监听器的回调

```java
Button button=(Button)this.findViewById(R.id.button);
button.setOnClickListener(newButton.OnClickListener(){
    //回调函数
```

```
@override
public void onClick(View v){
    buttonTextView.setText("按钮被点击了");
}
});
```

上面的代码给按钮加了一个事件监听器,自己不会显式地去调用 onClick()方法。用户触发了该按钮的点击事件后,它会由 Android 系统来自动调用。

应用场景二:Activity 生命周期中的回调

```
@Override
public void onCreate(BundlesaveInstanceState){
    super.onCreate(saveInstanceState);
    ...
}
@Override
public void onResume(){
    super.onResume();
    ...
}
```

上面的代码是创建 Activity 时,系统自带的 onCreate()方法,该方法不会被人为调用,但是它会在 Android 系统进行自动调用。

2. 获取版本号和安装 APK

在下载 APK 之前首先要获取到程序的本地版本号,当本地版本号与服务器版本号不一致时,弹出更新提醒对话框,进行下载安装。由于这部分代码功能比较独立,且在其他程序中也可以使用,因此将其抽取出来作为工具类 MyUtils,具体代码如【文件 1-3】所示。

【文件 1-3】MyUtils.java

```
1   public class MyUtils {
2   /**
3    * 获取版本号
4    * @param context
5    * @return 返回版本号
6    */
7   public static String getVersion(Context context){
8       //PackageManager 可以获取清单文件中的所有信息
9       PackageManager manager=context.getPackageManager();
10      try{
11          //getPackageName()获取到当前程序的包名
12          PackageInfo packageInfo=manager.getPackageInfo(
13          context.getPackageName(),0);
```

```
14          return packageInfo.versionName;
15      } catch (NameNotFoundException e){
16          e.printStackTrace();
17          return "";
18      }
19  }
20  /**
21   * 安装新版本
22   * @param activity
23   */
24  public static void installApk(Activity activity){
25      Intent intent=new Intent("android.intent.action.VIEW");
26      //添加默认分类
27      intent.addCategory("android.intent.category.DEFAULT");
28      //设置数据和类型
29      intent.setDataAndType(Uri.fromFile(new File("/mnt/sdcard/mobilesafe2.0.apk")),
30          "application/vnd.android.package-archive");
31      //如果开启的Activity退出时会回调当前Activity的onActivityResult
32      activity.startActivityForResult(intent,0);
33  }
34 }
```

代码说明：

- 第7~19行的getVersion()方法用于获取本地版本号，首先要获取到PackageManger对象，然后调用getPackageInfo()方法获取到PackageInfo对象，通过PackageInfo对象即可获取到本地版本号。
- 第24~33行的installApk()方法利用了隐式意图开启了系统中用于安装APK的Activity。

1.2.5 版本更新工具类

通过前面的讲解可知，欢迎界面需要获取服务器中程序的版本号,【实现版本号比对】→【弹出更新提醒对话框】→【弹出下载APK进度条】→【替换安装程序】等。由于这个逻辑是一个整体，因此将这块代码抽取出来放在工具类VersionUpdateUtils中，具体代码如【文件1-4】所示。

【文件1-4】VersionUpdateUtils.java

```
1 /** 更新提醒工具类 */
2 public class VersionUpdateUtils{
3     private static final int MESSAGE_NET_EEOR=101;
4     private static final int MESSAGE_IO_EEOR=102;
5     private static final int MESSAGE_JSON_EEOR=103;
6     private static final int MESSAGE_SHOEW_DIALOG=104;
7     protected static final int MESSAGE_ENTERHOME=105;
```

```java
8   /** 用于更新UI */
9   private Handler handler=new Handler(){
10      public void handleMessage(android.os.Message msg){
11          switch(msg.what){
12          case MESSAGE_IO_EEOR:
13              Toast.makeText(context,"IO异常",0).show();
14              enterHome();
15              break;
16          case MESSAGE_JSON_EEOR:
17              Toast.makeText(context,"JSON解析异常",0).show();
18              enterHome();
19              break;
20          case MESSAGE_NET_EEOR:
21              Toast.makeText(context,"网络异常",0).show();
22              enterHome();
23              break;
24          case MESSAGE_SHOEW_DIALOG:
25              showUpdateDialog(versionEntity);
26              break;
27          case MESSAGE_ENTERHOME:
28              Intent intent=new Intent(context,HomeActivity.class);
29              context.startActivity(intent);
30              context.finish();
31              break;
32          }
33      };
34  };
35  /** 本地版本号 */
36  private String mVersion;
37  private Activity context;
38  private ProgressDialog mProgressDialog;
39  private VersionEntity versionEntity;
40  public VersionUpdateUtils(String Version,Activity activity){
41      mVersion=Version;
42      context=activity;
43  }
44  /**
45   * 获取服务器版本号
46   */
```

```java
47  public void getCloudVersion(){
48      try {
49          HttpClient client=new DefaultHttpClient();
50          /*连接超时*/
51          HttpConnectionParams.setConnectionTimeout(client.getParams(),5000);
52          /*请求超时*/
53          HttpConnectionParams.setSoTimeout(client.getParams(),5000);
54          HttpGet httpGet=new HttpGet("http://172.16.25.14:8080/
55          updateinfo.html");
56          HttpResponse execute=client.execute(httpGet);
57          if(execute.getStatusLine().getStatusCode()==200){
58              //请求和响应都成功了
59              HttpEntity entity=execute.getEntity();
60              String result=EntityUtils.toString(entity,"gbk");
61              //创建jsonObject对象
62              JSONObject jsonObject=new JSONObject(result);
63              versionEntity=new VersionEntity();
64              String code=jsonObject.getString("code");
65              versionEntity.versioncode=code;
66              String des=jsonObject.getString("des");
67              versionEntity.description=des;
68              String apkurl=jsonObject.getString("apkurl");
69              versionEntity.apkurl=apkurl;
70              if(!mVersion.equals(versionEntity.versioncode)){
71                  //版本号不一致
72                  handler.sendEmptyMessage(MESSAGE_SHOEW_DIALOG);
73              }
74          }
75      } catch(ClientProtocolException e){
76          handler.sendEmptyMessage(MESSAGE_NET_EEOR);
77          e.printStackTrace();
78      } catch(IOException e){
79          handler.sendEmptyMessage(MESSAGE_IO_EEOR);
80          e.printStackTrace();
81      } catch(JSONException e){
82          handler.sendEmptyMessage(MESSAGE_JSON_EEOR);
83          e.printStackTrace();
84      }
85  }
```

```java
86  /**
87   * 弹出更新提示对话框
88   * @param versionEntity
89   */
90  private void showUpdateDialog(final VersionEntity versionEntity){
91      //创建dialog
92      AlertDialog.Builder builder=new Builder(context);
93      builder.setTitle("检查到新版本："+versionEntity.versioncode);//设置标题
94      builder.setMessage(versionEntity.description);
95      //根据服务器返回描述,设置升级描述信息
96      builder.setCancelable(false);  //设置不能点击手机返回按钮隐藏对话框
97      builder.setIcon(R.drawable.ic_launcher);//设置对话框图标
98      //设置立即升级按钮点击事件
99      builder.setPositiveButton("立即升级",new DialogInterface.OnClick
100     Listener(){
101         @Override
102         public void onClick(DialogInterface dialog,int which){
103             initProgressDialog();
104             downloadNewApk(versionEntity.apkurl);
105         }
106     });
107     // 设置暂不升级按钮点击事件
108     builder.setNegativeButton("暂不升级",new DialogInterface.OnClick
109     Listener(){
110         @Override
111         public void onClick(DialogInterface dialog,int which){
112             dialog.dismiss();
113             enterHome();
114         }
115     });
116     builder.show();  //对话框必须调用show方法否则不显示
117 }
118 /**
119  * 初始化进度条对话框
120  */
121 private void initProgressDialog(){
122     mProgressDialog = new ProgressDialog(context);
123     mProgressDialog.setMessage("准备下载...");
124     mProgressDialog.setProgressStyle(ProgressDialog.STYLE_HORIZONTAL);
```

```
125        mProgressDialog.show();
126    }
127    /**
128     * 下载新版本
129     */
130    protected void downloadNewApk(String apkurl){
131        DownLoadUtils downLoadUtils=new DownLoadUtils();
132        downLoadUtils.downapk(apkurl,"/mnt/sdcard/mobilesafe2.0.apk",
133        new MyCallBack(){
134            @Override
135            public void onSuccess(ResponseInfo<File> arg0){
136                //TODO Auto-generated method stub
137                mProgressDialog.dismiss();
138                MyUtils.installApk(context);
139            }
140            @Override
141            public void onLoadding(long total,long current,boolean isUploading){
142                //TODO Auto-generated method stub
143                mProgressDialog.setMax((int)total);
144                mProgressDialog.setMessage("正在下载...");
145                mProgressDialog.setProgress((int) current);
146            }
147            @Override
148            public void onFailure(HttpException arg0,String arg1) {
149                //TODO Auto-generated method stub
150                mProgressDialog.setMessage("下载失败");
151                mProgressDialog.dismiss();
152                enterHome();
153            }
154        });
155    }
156    private void enterHome(){
157        handler.sendEmptyMessageDelayed(MESSAGE_ENTERHOME, 2000);
158    }
159}
```

代码说明：

- 第9~34行的Handler代码用于线程间通信，及时通知主线程更新UI，并进入主界面。
- 第47~85行的getCloudVersion()方法用于访问网络获取服务器版本号，首先创建一个HttpClient对象，然后通过HttpConnectionParams设置链接超时时间和请求超时时间，并通过 HttpGet 请求 updateinfo.html 页面，解析该页面的 JSON 数据，与本地版本号进行比对，如果不一致则使用 Handler 发送消息。
- 第 90~117 行的 showUpdateDialog 方法用于弹出升级对话框，当点击"暂不升级"按钮时，会进入主界面，当点击"立即升级"时，会初始化下载对话框调用下载 APK 的方法。
- 第 130~155 行的 downloadNewApk()方法用于下载 APK，在该方法中调用了 DownLoadUtils.downapk()方法。当 APK 下载完成后，调用了 MyUtils.installApk()方法进行安装。

上述代码是一个完整的下载更新流程，因此只需要在 Activity 中创建 VersionUpdateUtils 实例，并在子线程中调用 getCloudVersion()方法即可。

1.2.6 版本信息的实体类

由于从服务器中获取的程序版本号、版本描述、下载地址需要存储到实体类中，因此需要定义一个实体类 VersionEntity，具体代码如【文件 1-5】所示。

【文件 1-5】VersionEntity.java

```
1  public class VersionEntity{
2      /**服务器版本号*/
3      public String versioncode;
4      /**版本描述*/
5      public String description;
6      /**apk下载地址*/
7      public String apkurl;
8  }
```

1.2.7 欢迎界面逻辑

创建好一系列的工具类和实体类之后，接下来需要在 SplashActivity 中调用相应的代码，在欢迎页面中实现版本更新操作，具体代码如【文件 1-6】所示。

【文件 1-6】SplashActivity.java

```
1  public class SplashActivity extends Activity{
2      /** 应用版本号 */
3      private TextView mVersionTV;
4      /** 本地版本号 */
5      private String mVersion;
6      @Override
7      protected void onCreate(Bundle savedInstanceState){
8          super.onCreate(savedInstanceState);
9          //设置该Activity没有标题栏，在加载布局之前调用
10         requestWindowFeature(Window.FEATURE_NO_TITLE);
11         setContentView(R.layout.activity_splash);
12         mVersion=MyUtils.getVersion(getApplicationContext());
13         initView();
```

```
14    final VersionUpdateUtils updateUtils=new VersionUpdateUtils
15    (mVersion,SplashActivity.this);
16    new Thread(){
17        public void run(){
18            //获取服务器版本号
19            updateUtils.getCloudVersion();
20        };
21    }.start();
22  }
23  /** 初始化控件 */
24  private void initView(){
25      mVersionTV=(TextView) findViewById(R.id.tv_splash_version);
26      mVersionTV.setText("版本号 "+mVersion);
27  }
28 }
```

代码说明：
- 第 12 行代码通过 MyUtils 类中的 getVersion()方法获取到应用的本地版本号。
- 第 14～21 行代码创建 VersionUpdateUtils 对象并开启子线程用于获取服务器端程序版本号。
- 第 24～27 行 initView()方法用于将获取到的本地版本号显示在界面上。

由于本程序需要请求网络下载 APK 到手机内存卡，因此需要在 AndroidManifest.xml 文件中添加访问网络和写 SD 卡的权限，具体代码如下所示：

```
<uses-permission android:name="android.permission.INTERNET"/>
<uses-permission android:name="android.permission.WRITE_EXTERNAL_STORAGE"/>
```

1.3 主　界　面

当打开一个应用时，首先会显示欢迎界面，然后跳转到程序主界面。主界面的作用就是展示当前应用的基本功能以及界面风格。本小节将针对程序主界面的开发进行详细讲解。

1.3.1 主界面 UI

从功能展示界面可以看出，主界面分为上下两部分：上部分放置一个安仔图标以及一行文字；下部分用 GridView（九宫格的形式）放置 9 个功能图标及功能文字。主界面的图形化界面如图 1-17 所示。

图 1-17 所示主界面对应的布局文件如【文件 1-7】所示。

【文件 1-7】activity_home.xml

```
<?xml version="1.0" encoding="utf-8"?>
<LinearLayout xmlns:android="http://schemas.android.
    com/apk/res/android"
    android:layout_width="match_parent"
```

图 1-17　主界面

```xml
    android:layout_height="match_parent"
    android:layout_gravity="center"
    android:background="@drawable/bg_home"
    android:orientation="vertical">
    <LinearLayout
        android:id="@+id/ll"
        android:layout_width="match_parent"
        android:layout_height="wrap_content"
        android:layout_marginTop="10dp"
        android:gravity="center"
        android:orientation="horizontal">
        <ImageView
            android:layout_width="100dp"
            android:layout_height="107.5dp"
            android:background="@drawable/superman"/>
        <TextView
            android:id="@+id/tv_ad"
            android:layout_width="wrap_content"
            android:layout_height="wrap_content"
            android:gravity="center"
            android:paddingTop="30dp"
            android:text="主银，我是您的手机小护卫"
            android:textColor="#90000000"
            android:textScaleX="1.1"
            android:textSize="16sp"
            android:typeface="normal"/>
    </LinearLayout>
    <RelativeLayout
        android:layout_width="match_parent"
        android:layout_height="match_parent">
        <GridView
            android:id="@+id/gv_home"
            android:layout_width="match_parent"
            android:layout_height="wrap_content"
            android:layout_centerInParent="true"
            android:layout_marginLeft="10dp"
            android:layout_marginRight="10dp"
            android:cacheColorHint="#00000000"
            android:gravity="center"
            android:horizontalSpacing="10dp"
```

```
                android:numColumns="3"
                android:verticalSpacing="10dp">
        </GridView>
    </RelativeLayout>
</LinearLayout>
```

在上述布局文件中,使用了一个 GridView 控件,该控件是一个二维的 ViewGroup,通过它可以在界面中展示九宫格效果。

1.3.2 主界面 Item 布局

由于主界面的 GridView 有 9 个 Item,每个 Item 的格式都是一样的,都由一张图片和一个标题组成,因此可以创建一个 Item 布局为 GridView 设置每个条目数据,Item 布局的图形化界面如图 1-18 所示。

图 1-18 中 Item 所对应的布局文件如【文件 1-8】所示。

【文件 1-8】item_home.xml

图 1-18 Item 布局

```xml
<?xml version="1.0" encoding="utf-8"?>
<LinearLayout xmlns:android="http://schemas.android.com/
    apk/res/android"
    android:layout_width="wrap_content"
    android:layout_height="wrap_content"
    android:gravity="center"
    android:orientation="vertical">
    <ImageView
        android:id="@+id/iv_icon"
        android:layout_width="80dp"
        android:layout_height="80dp"
        android:background="@drawable/atools"/>
    <TextView
        android:id="@+id/tv_name"
        android:layout_width="wrap_content"
        android:layout_height="wrap_content"
        android:text="高级工具"
        android:layout_marginTop="3dp"
        android:textColor="#80000000"
        android:textSize="18sp"/>
</LinearLayout>
```

1.3.3 数据适配器

主界面的 Item 布局创建好之后,要想将这个 Item 中的功能图标以及文字填充到主界面中,还需要创建一个布局填充器 HomeAdapter,具体代码如【文件 1-9】所示。

【文件 1-9】 HomeAdapter.java

```java
/**GridView 显示界面用到的 Adapter*/
public class HomeAdapter extends BaseAdapter{
    int[] imageId={R.drawable.safe, R.drawable.callmsgsafe, R.drawable.app,
                R.drawable.trojan, R.drawable.sysoptimize,R.drawable. taskmanager,
                R.drawable.netmanager,R.drawable.atools,R.drawable.
                settings};
    String[] names={"手机防盗","通讯卫士","软件管家","手机杀毒","缓存清理",
                "进程管理","流量统计","高级工具","设置中心" };
    private Context context;
    public HomeAdapter(Context context){
        this.context=context;
    }
    //设置 gridView 一共有多少个条目
    @Override
    public int getCount(){
        return 9;
    }
    //设置每个条目的界面
    @Override
    public View getView(int position,View convertView,ViewGroup parent){
        View view=View.inflate(context,R.layout.item_home, null);
        ImageView iv_icon=(ImageView) view.findViewById(R.id.iv_icon);
        TextView tv_name=(TextView) view.findViewById(R.id.tv_name);
        iv_icon.setImageResource(imageId[position]);
        tv_name.setText(names[position]);
        return view;
    }
    //后面两个方法暂时不需要设置
    @Override
    public Object getItem(int position){
        return null;
    }
    @Override
    public long getItemId(int position){
        return 0;
    }
}
```

1.3.4 主界面逻辑

主界面中的 Item 布局以及 HomeAdapter 都已实现，接下来编写对应的 HomeActivity。该程序用于加载主界面布局，并将功能图标及文字展示到主界面中，同时将跳转页面的点击事件实现。具体代码如【文件 1-10】所示。

【文件 1-10】HomeActivty.java

```java
1  public class HomeActivity extends Activity{
2      private long mExitTime;
3      private GridView gv_home;
4      @Override
5      protected void onCreate(Bundle savedInstanceState){
6          super.onCreate(savedInstanceState);
7          //初始化布局
8          requestWindowFeature(Window.FEATURE_NO_TITLE);
9          setContentView(R.layout.activity_home);
10         //初始化GridView
11         gv_home=(GridView) findViewById(R.id.gv_home);
12         gv_home.setAdapter(new HomeAdapter(HomeActivity.this));
13         //设置条目的点击事件
14         gv_home.setOnItemClickListener(new OnItemClickListener(){
15             //parent 代表 gridView, view 代表每个条目的 view 对象, postion 代表每个条目的位置
16             @Override
17             public void onItemClick(AdapterView<?> parent,View view,
18                 int position,long id){
19                 switch(position){
20                     case 0: //手机防盗
21                         if(isSetUpPassword()){
22                             //弹出输入密码对话框
23                             showInterPswdDialog();
24                         } else {
25                             //弹出设置密码对话框
26                             showSetUpPswdDialog();
27                         }
28                         break;
29                     case 1: //通讯卫士
30                         startActivity(SecurityPhoneActivity.class);
31                         break;
32                     case 2: //软件管家
33                         startActivity(AppManagerActivity.class);
```

```java
34                     break;
35                 case 3: //病毒查杀
36                     startActivity(VirusScanActivity.class);
37                     break;
38                 case 4: //缓存清理
39                     startActivity(CacheClearListActivity.class);
40                     break;
41                 case 5: //进程管理
42                     startActivity(ProcessManagerActivity.class);
43                     break;
44                 case 6: //流量统计
45                     startActivity(TrafficMonitoringActivity.class);
46                     break;
47                 case 7: //高级工具
48                     startActivity(AdvancedToolsActivity.class);
49                     break;
50                 case 8: //设置中心
51                     startActivity(SettingsActivity.class);
52                     break;
53             }
54         }
55     });
56 }
57 /**
58  * 开启新的Activity不关闭自己
59  * @param cls，新的Activity的字节码
60  */
61 public void startActivity(Class<?> cls){
62     Intent intent=new Intent(HomeActivity.this,cls);
63     startActivity(intent);
64 }
65 /**
66  * 按两次返回键退出程序
67  */
68 @Override
69 public boolean onKeyDown(int keyCode, KeyEvent event){
70     if(keyCode==KeyEvent.KEYCODE_BACK){
71         if((System.currentTimeMillis()-mExitTime)<2000){
72             System.exit(0);
```

```
73          }else{
74              Toast.makeText(this,"再按一次退出程序",Toast.LENGTH_SHORT).show();
75              mExitTime=System.currentTimeMillis();
76          }
77          return true;
78      }
79      return super.onKeyDown(keyCode,event);
80  }
81 }
```

代码说明：

- 第 11~12 行代码获取 GridView 控件，并为该控件设置 Adapter 填充数据。
- 第 14~55 行代码为 GridView 控件注册条目点击的监听事件，当点击每个条目时，相应的模块会响应点击事件。暂不使用的功能可以将其注释，这里只是为了方便大家观察代码。
- 第 61~64 行的 startActivity()方法用于开启新的 Activity，在点击每个功能图标进行界面跳转时调用。将这段代码单独作为一个方法，可以大大简化代码量，在页面跳转时直接传入相应的 Activity 即可。
- 第 69~80 行的 onKeyDown()方法用于实现按两次返回键退出程序，在 Android 应用中经常要判断用户对返回键的操作，一般为了防止误操作，都是在用户连续按两次返回键时才退出程序。通常情况下，实现该功能时先记录上次点击的时间，然后与本次点击的时间进行比较，当两次时间间隔小于某个值时，退出程序，否则提示"再按一次退出程序"，同时更新上次点击时间。

注意：在 Android 系统中，四大组件在使用时都需要在清单文件中进行注册，考虑到代码量的问题，在后面开发中关于 Activity 注册的代码将被省略，大家在编程时自行添加即可。

本 章 小 结

本章首先针对手机安全卫士项目进行分析、代码结构介绍，然后针对欢迎界面以及主界面开发进行详细讲解。本章所讲的欢迎界面以及主界面知识，在任何一个项目中都会使用，因此大家一定要熟练掌握其原理，并能通过代码实现。

【面试精选】
1. 请问 Android 程序的真正入口是什么？
2. 请问 JSON 数据与 XML 数据各有哪些优缺点？
扫描右方二维码，查看面试题答案！

第 2 章

➡ 手机防盗模块

学习目标
- 了解手机防盗模块功能；
- 掌握防盗密码功能的开发；
- 掌握设置向导功能的开发；
- 掌握手势识别器的使用。

在日常生活中，难免会发生手机丢失的情况，此时如何才能不让别人使用自己的手机并找回呢？为此，我们开发了手机防盗模块，该模块可以定位手机、远程锁屏、清除数据等。本章将针对手机防盗模块进行详细讲解。

2.1 模 块 概 述

2.1.1 功能介绍

手机防盗模块主要用于 SIM 卡变更提醒、GPS 定位、远程锁定手机、远程删除数据等。接下来针对该模块功能进行介绍，具体如下：

1. 设置密码

手机防盗功能是用户的一个隐私设置，因此当第一次进入"手机防盗"模块时需要设置密码，密码设置成功后当再次进入时需要输入密码，才能进入"手机防盗"模块并修改当前设置，具体如图 2-1 所示。

图 2-1　手机防盗界面

2. 设置向导

设置向导主要有四个界面，分别用来展示当前模块的功能、SIM 卡绑定情况、选择安全联系人、设置完成界面，这几个界面都是通过手指滑动屏幕来实现界面切换的。

当进入 SIM 卡绑定界面，点击"绑定 SIM 卡"按钮后，滑动屏幕进入选择安全联系人界面，可以选择通讯录中的号码作为安全号码或者直接输入安全号码，当 SIM 卡变更时会向安全号码发送报警短信。当前功能都设置完成后，滑动屏幕进入设置完成界面，此时显示防盗保护已经开启，具体如图 2-2 所示。

图 2-2 设置向导界面

3. 联系人列表

在设置向导中添加安全联系人时，点击"请输入安全号码"右侧的加号按钮，还可以跳转到联系人列表界面，从该列表中选取安全联系人，具体如图 2-3 所示。

4. 手机防盗界面

手机防盗界面会显示当前设置的安全号码、防盗保护是否开启、重新进入设置向导以及短信指令功能介绍。所谓短信指令就是编辑好的一些特殊指令，当向手机发送对应的指令时会进行不同的操作，例如向手机发送短信"#*alarm*#"此时手机会播放报警音乐，界面如图 2-4 所示。

图 2-3 联系人列表界面　　图 2-4 手机防盗界面

至此，手机防盗功能已经介绍完成，编程者在开发该功能界面时可以参考上述效果图。

2.1.2 开发流程图

在开发手机防盗模块之前，可以先画一张流程图，了解程序逻辑方便程序开发。手机防盗模块的流程图如图 2-5 所示。

从图 2-5 可以看出，当点击"手机防盗"功能图标时，会判断是否设置过防盗密码，如果已经设置，则弹出输入密码对话框；否则弹出设置密码对话框，密码设置成功后会再次弹出输入密码对话框，当密码输入正确时会进入手机防盗界面，然后判断是否设置过防盗向导。如果设置过，则展示手机防盗界面功能，否则进入设置向导界面依次进行设置，设置完成后进入手机防盗界面。

2.1.3 代码结构

手机防盗模块中涉及的代码比较多，逻辑也比较复杂，为了让编程者对该模块的结构更加清楚，接下来通过一个图例来展示手机防盗模块的代码结构，如图 2-6 所示。

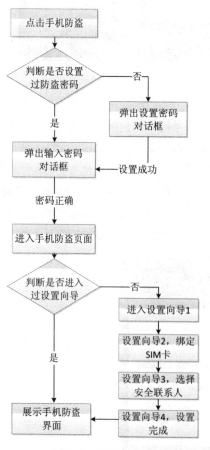

图 2-5 手机防盗流程图

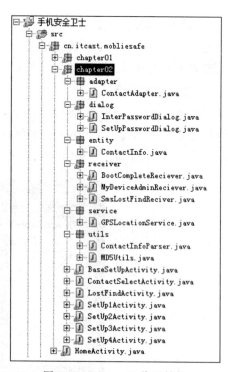

图 2-6 chapter02 代码结构

下面按照结构顺序依次介绍 chapter02 包中的文件，具体如下：

- ContactAdapter.java：用于填充联系人界面的数据适配器；
- InterPasswordDialog.java：自定义的 Dialog 对话框，用于输入防盗密码；
- SetUpPasswordDialog.java：自定义的 Dialog 对话框，用于设置防盗密码；
- ContactInfo.java：用于存储联系人信息的实体类，其中包括 id、name、phone 字段；
- BootCompleteReciever.java：用于监听手机开机启动的广播接收者，检测 SIM 卡是否更换；
- MyDeviceAdminReceiver.java：用于获取超级管理员权限的广播接收者；
- SmsLostFindReceiver.java：监听短信到来的广播接收者，接收到指令后进行相应的报警操作；
- GPSLocationService.java：用于定位手机的服务；
- ContactInfoParser.java：用于从数据库中解析联系人信息；
- MD5Utils.java：加密算法的工具类；
- BaseSetUpActivity.java：设置向导的公共父类，封装了一些手势识别的方法；
- ContactSelectActivity.java：获取手机中的联系人并展示到界面中；
- LostFindActivity.java：用于显示手机防盗界面以及防盗指令；
- SetUp1Activity.java：显示设置向导第一个界面；
- SetUp2Activity.java：显示设置向导第二个界面并绑定 SIM 卡；
- SetUp3Activity.java：显示设置向导第三个界面并选择安全联系人；
- SetUp4Activity.java：显示设置向导第四个界面并开启防盗保护功能。

2.2 设置密码

手机防盗模块对于用户来说是一个较为隐私的功能，不能让其他人随意查看，因此必须要设置一个密码来对该模块进行保护。本节将针对设置密码功能进行详细讲解。

2.2.1 设置密码界面

设置防盗密码是进入手机防盗模块的一个入口，两次密码输入一致时，程序会自动保存输入的密码并弹出输入密码对话框。设置防盗密码界面布局主要有两个 EditText 和两个 Button 按钮，以及用于显示线条的 View 控件，设置密码对话框的图形化界面如图 2-7 所示。

图 2-7 所示设置密码对话框的布局文件如【文件 2-1】所示。

【文件 2-1】 setup_password_dialog.xml

图 2-7 设置密码对话框

```
<?xml version="1.0" encoding="utf-8"?>
<LinearLayout xmlns:android="http://schemas.android.
    com/apk/res/android"
    android:layout_width="260dp"
    android:layout_height="220dp"
    android:minHeight="150dp"
```

```xml
android:orientation="vertical"
android:layout_gravity="center"
android:background="@drawable/coner_bg_white" >
<TextView
    android:id="@+id/tv_setuppwd_title"
    android:layout_width="match_parent"
    android:layout_height="wrap_content"
    android:padding="8dp"
    android:layout_margin="5dp"
    android:textColor="#1ABDE6"
    android:text="设置密码"
    android:textSize="20sp"/>
<View
    android:layout_width="match_parent"
    android:layout_height="1.5px"
    android:background="#1ABDE6"
    android:layout_marginBottom="10dp"/>
<EditText
    android:id="@+id/et_firstpwd"
    android:layout_margin="5dp"
    android:layout_width="match_parent"
    android:layout_height="40dp"
    android:inputType="textPassword"
    android:hint="请输入密码"
    android:background="@drawable/edit_normal"/>
<EditText
    android:id="@+id/et_affirm_password"
    android:layout_margin="5dp"
    android:layout_width="match_parent"
    android:layout_height="40dp"
    android:inputType="textPassword"
    android:hint="请再次输入密码"
    android:background="@drawable/edit_normal"/>
<View
    android:layout_marginTop="10dp"
    android:layout_width="match_parent"
    android:layout_height="1.0px"
    android:background="#1ABDE6"/>
<LinearLayout
```

```
    android:layout_width="match_parent"
    android:layout_height="match_parent"
    android:orientation="horizontal"
    android:gravity="center">
    <Button
        android:id="@+id/btn_ok"
        android:layout_width="0dp"
        android:layout_height="wrap_content"
        android:layout_weight="1"
        android:text="确认"
        android:background="@android:color/white"/>
    <View
    android:layout_width="1.0px"
    android:layout_height="match_parent"
    android:background="#1ABDE6"/>
    <Button
        android:id="@+id/btn_cancle"
        android:layout_width="0dp"
        android:layout_height="wrap_content"
        android:layout_weight="1"
        android:text="取消"
        android:background="@android:color/white"/>
    </LinearLayout>
</LinearLayout>
```

由于设置密码对话框和输入密码对话框的布局风格一致，因此可以使用 shape 属性自定义对话框的样式。在 drawable 目录中创建一个 xml 文件，设置标签为 shape，在该文件中定义对话框的圆角以及颜色，并通过 "android:background="@drawable/coner_bg_white"" 将自定义属性设置到 LinearLayout 上。设置密码对话框的自定义属性如【文件 2-2】所示。

【文件 2-2】res/drawable/coner_bg_white.xml

```
<?xml version="1.0" encoding="utf-8"?>
<shape xmlns:android="http://schemas.android.com/apk/res/android" >
    <corners android:radius="6.0dp"/>
    <solid android:color="#ffffff"/>
</shape>
```

上述文件中的<shape>标签用于设置形状，它有六个子标签，分别为<corners>、<solid>、<gradient>、<padding>、<size>、<stroke>。其中<corners>用于表示圆角，android:radius="6.0dp" 属性用于指定圆角的半径。<solid>用于指定填充颜色，android:color="#ffffff"表示颜色为白色。

2.2.2 自定义对话框样式

在 Android 系统中,自带的对话框通常是一个灰色的并带有标题栏和按钮,由于这种对话框样式在当前界面使用并不美观,因此需要自定义对话框样式。具体代码如【文件 2-3】所示。

【文件 2-3】res/values/styles.xml

```xml
<!-- 自定义对话框样式 -->
<style name="dialog_custom" parent="android:style/Theme.Dialog">
    <item name="android:windowFrame">@null</item>
    <item name="android:windowNoTitle">true</item>
    <item name="android:background">#00000000</item>
    <item name="android:windowBackground">@android:color/transparent</item>
</style>
```

上述代码中,第一个 Item 表示去掉边框,第二个 Item 表示无标题栏,第三个表示背景颜色为黑色,第四个 Item 表示对话框背景透明。

2.2.3 设置密码逻辑

设置防盗密码对话框的 UI 编写完成后,接下来编写设置密码对话框的逻辑类,该类继承自 Dialog,主要用于初始化控件以及响应按钮的点击事件。具体代码如【文件 2-4】所示。

【文件 2-4】SetUpPasswordDialog.java

```java
1  public class SetUpPasswordDialog extends Dialog implements
2      android.view.View.OnClickListener{
3   /**标题栏*/
4   private TextView mTitleTV;
5   /**首次输入密码文本框*/
6   public EditText mFirstPWDET;
7   /**确认密码文本框*/
8   public EditText mAffirmET;
9   /**回调接口*/
10  private MyCallBack myCallBack;
11  public SetUpPasswordDialog(Context context){
12      super(context,R.style.dialog_custom); // 引入自定义对话框样式
13  }
14  public void setCallBack(MyCallBack myCallBack){
15      this.myCallBack = myCallBack;
16  }
17  @Override
18  protected void onCreate(Bundle savedInstanceState){
19      setContentView(R.layout.setup_password_dialog);
```

```java
20        super.onCreate(savedInstanceState);
21        initView();
22    }
23    /**初始化控件*/
24    private void initView(){
25        mTitleTV=(TextView) findViewById(R.id.tv_interpwd_title);
26        mFirstPWDET=(EditText) findViewById(R.id.et_firstpwd);
27        mAffirmET=(EditText) findViewById(R.id.et_affirm_password);
28        findViewById(R.id.btn_ok).setOnClickListener(this);
29        findViewById(R.id.btn_cancle).setOnClickListener(this);
30    }
31    /**
32     * 设置对话框标题栏
33     * @param title
34     */
35    public void setTitle(String title){
36        if(!TextUtils.isEmpty(title)){
37            mTitleTV.setText(title);
38        }
39    }
40    @Override
41    public void onClick(View v){
42        switch (v.getId()){
43            case R.id.btn_ok:
44                myCallBack.ok();
45                break;
46            case R.id.btn_cancle:
47                myCallBack.cancle();
48                break;
49        }
50    }
51    public interface MyCallBack{
52        void ok();
53        void cancle();
54    }
55 }
```

代码说明：

- 第 11～13 行的 SetUpPasswordDialog()方法为构造方法，在该方法中通过 super(context, R.style.dialog_custom)引入自定义对话框样式 dialog_custom。

- 第 14～16 行的 setCallBack()方法定义了一个回调函数 setCallBack(MyCallBack myCallBack) 传递一个 MyCallBack 接口；
- 第 35～39 行的 setTitle()方法用于设置输入密码对话框的标题栏；
- 第 41～50 行的 onClick()方法用于响应按钮的点击事件,当点击确定按钮时调用 MyCallBack 中的 ok()方法,点击取消按钮时调用 cancle()按钮；
- 第 51～54 行定义一个 MyCallBack 接口,该接口中有两个方法分别用于处理按钮的确定与取消。

2.2.4 MD5 加密算法

为了保证账户安全,在保存用户密码时,通常会采用 MD5 加密算法,这种算法是不可逆的,具有一定的安全性。由于加密算法功能比较独立,因此将其抽取出作为工具类单独存放,具体代码如【文件 2-5】所示。

【文件 2-5】MD5Utils.java

```
1  public class MD5Utils{
2    /**
3     * md5 加密的算法
4     * @param text
5     * @return
6     */
7    public static String encode(String text){
8      try{
9        MessageDigest digest=MessageDigest.getInstance("md5");
10       byte[] result=digest.digest(text.getBytes());
11       StringBuilder sb=new StringBuilder();
12       for(byte b:result){
13         int number=b&0xff;
14         String hex=Integer.toHexString(number);
15         if(hex.length()==1){
16           sb.append("0"+hex);
17         }else{
18           sb.append(hex);
19         }
20       }
21       return sb.toString();
22     }catch (NoSuchAlgorithmException e){
23       e.printStackTrace();
24       //can't reach
25       return "";
26     }
```

```
27    }
28 }
```

至此，设置防盗密码的 UI 以及基本逻辑代码已经完成，在 HomeActivity 中进行调用即可。

2.3 输入密码

输入防盗密码功能相对来说比较简单，用户输入密码后点击"确认"按钮，程序会与当初设置的密码进行比对，密码一致则进入设置向导界面。本节将针对输入密码功能进行详细讲解。

2.3.1 输入密码界面

输入防盗密码布局与设置密码布局类似，只不过输入密码布局只有一个 EditText 而已，输入密码对话框的图形化界面如图 2-8 所示。

图 2-8 所示输入密码对话框的布局文件如【文件 2-6】所示。

【文件 2-6】inter_password_dialog.xml

图 2-8 输入密码对话框

```xml
<?xml version="1.0" encoding="utf-8"?>
<LinearLayout xmlns:android="http://schemas.android.com/apk/res/android"
    android:layout_width="260dp"
    android:layout_height="170dp"
    android:minHeight="150dp"
    android:orientation="vertical"
    android:layout_gravity="center"
    android:background="@drawable/coner_bg_white" >
    <TextView
        android:id="@+id/tv_interpwd_title"
        android:layout_width="match_parent"
        android:layout_height="wrap_content"
        android:padding="8dp"
        android:layout_margin="5dp"
        android:textColor="#1ABDE6"
        android:text="输入密码"
        android:textSize="20sp"/>
    <View
        android:layout_width="match_parent"
        android:layout_height="1.5px"
        android:background="#1ABDE6"
        android:layout_marginBottom="10dp"/>
    <EditText
```

```xml
        android:id="@+id/et_inter_password"
        android:layout_margin="5dp"
        android:layout_width="match_parent"
        android:layout_height="40dp"
        android:inputType="textPassword"
        android:hint="请输入密码"
        android:background="@drawable/edit_normal"/>
    <View
        android:layout_marginTop="10dp"
        android:layout_width="match_parent"
        android:layout_height="1.0px"
        android:background="#1ABDE6"/>
    <LinearLayout
        android:layout_width="match_parent"
        android:layout_height="match_parent"
        android:orientation="horizontal"
        android:gravity="center">
        <Button
            android:id="@+id/btn_comfirm"
            android:layout_width="0dp"
            android:layout_height="wrap_content"
            android:layout_weight="1"
            android:text="确认"
            android:background="@android:color/white"/>
        <View
            android:layout_width="1.0px"
            android:layout_height="match_parent"
            android:background="#1ABDE6"/>
        <Button
            android:id="@+id/btn_dismiss"
            android:layout_width="0dp"
            android:layout_height="wrap_content"
            android:layout_weight="1"
            android:text="取消"
            android:background="@android:color/white"/>
    </LinearLayout>
</LinearLayout>
```

2.3.2 输入密码逻辑

输入防盗密码对话框的 UI 界面编写完成后，下面编写输入密码对话框的逻辑类，该类与设置密码非常相似，这里不再过多介绍。具体代码如【文件 2-7】所示。

【文件 2-7】InterPasswordDialog.java

```java
public class InterPasswordDialog extends Dialog implements
android.view.View.OnClickListener{
    /**对话框标题*/
    private TextView mTitleTV;
    /**输入密码文本框*/
    private EditText mInterET;
    /**确认按钮*/
    private Button mOKBtn;
    /**取消按钮*/
    private Button mCancleBtn;
    private MyCallBack myCallBack;
    private Context context;
    public InterPasswordDialog(Context context) {
        super(context,R.style.dialog_custom);
        this.context = context;
    }
    public void setCallBack(MyCallBack myCallBack){
        this.myCallBack = myCallBack;
    }
    @Override
    protected void onCreate(Bundle savedInstanceState) {
        setContentView(R.layout.inter_password_dialog);
        super.onCreate(savedInstanceState);
        initView();
    }
    private void initView() {
        mTitleTV = (TextView) findViewById(R.id.tv_interpwd_title);
        mInterET = (EditText) findViewById(R.id.et_inter_password);
        mOKBtn = (Button) findViewById(R.id.btn_comfirm);
        mCancleBtn = (Button) findViewById(R.id.btn_dismiss);
        mOKBtn.setOnClickListener(this);
        mCancleBtn.setOnClickListener(this);
    }
    public void setTitle(String title){
        if(!TextUtils.isEmpty(title)){
```

```
36          mTitleTV.setText(title);
37      }
38  }
39  public String getPassword(){
40      return mInterET.getText().toString();
41  }
42  @Override
43  public void onClick(View v) {
44      switch (v.getId()) {
45      case R.id.btn_comfirm:
46          myCallBack.confirm();
47          break;
48      case R.id.btn_dismiss:
49          myCallBack.cancle();
50          break;
51      }
52  }
53  public interface MyCallBack{
54      void confirm();
55      void cancle();
56  }
57 }
```

2.4 设置向导界面

设置向导中有四个 UI 界面，分别为手机防盗模块的展示界面、绑定 SIM 卡界面、选择安全联系人界面、设置完成界面。本节将针对这四个界面以及涉及的自定义样式和属性分别进行详细讲解。

2.4.1 小圆点界面

设置向导的四个界面中，都会用到显示滑动到当前界面的小圆点 ●●●●。因此先来开发小圆点界面，方便其他界面使用，然后依次开发设置向导界面。

界面中这些小圆点实际上就是 RadioGroup 中放了一组 RadioButton，通过 android:background 属性来指定按钮的颜色，具体代码如【文件 2-8】所示。

【文件 2-8】setup_radiogroup.xml

```
<?xml version="1.0" encoding="utf-8"?>
<LinearLayout xmlns:android="http://schemas.android.com/apk/res/android"
    android:layout_width="match_parent"
    android:layout_height="wrap_content"
    android:gravity="center"
```

```xml
        android:orientation="vertical" >
    <RadioGroup
        android:layout_gravity="center"
        android:layout_width="wrap_content"
        android:layout_height="wrap_content"
        android:orientation="horizontal">
        <RadioButton
            android:id="@+id/rb_first"
            android:layout_width="10dp"
            android:layout_height="10dp"
            android:button="@null"
            android:background="@drawable/circle_purple_bg_selector"/>
        <RadioButton
            android:id="@+id/rb_second"
            android:layout_width="10dp"
            android:layout_height="10dp"
            android:button="@null"
            android:layout_marginLeft="15dp"
            android:background="@drawable/circle_purple_bg_selector"/>
        <RadioButton
            android:id="@+id/rb_third"
            android:layout_width="10dp"
            android:layout_height="10dp"
            android:button="@null"
            android:layout_marginLeft="15dp"
            android:background="@drawable/circle_purple_bg_selector"/>
        <RadioButton
            android:id="@+id/rb_four"
            android:layout_width="10dp"
            android:layout_height="10dp"
            android:button="@null"
            android:layout_marginLeft="15dp"
            android:background="@drawable/circle_purple_bg_selector"/>
    </RadioGroup>
</LinearLayout>
```

在上述代码中，使用了一个背景选择器 circle_purple_bg_selector，该背景选择器用于控制按钮点击时显示的颜色，当滑动到第几个页面时所对应的第几个按钮就显示为紫色，其他按钮显示白色，该背景选择器代码如【文件2-9】所示。

【文件 2-9】res/drawable/circle_purple_bg_selector.xml

```xml
<?xml version="1.0" encoding="utf-8"?>
<selector xmlns:android="http://schemas.android.com/apk/res/android" >
    <item android:state_checked="true" android:drawable="@drawable/circle_purple"/>
    <item android:state_checked="false" android:drawable="@drawable/circle_white"/>
</selector>
```

从上述代码可以看出，circle_purple_bg_selector.xml 并没有直接指定颜色，而是引入了另外两个自定义属性 circle_purple.xml 和 circle_white.xml，这两个自定义属性中分别指定了按钮颜色为紫色、白色，按钮形状为实心圆，具体代码如【文件 2-10】和【文件 2-11】所示。

【文件 2-10】res/drawable/circle_purple.xml

```xml
<?xml version="1.0" encoding="utf-8"?>
<shape xmlns:android="http://schemas.android.com/apk/res/android" android:shape="oval">
    <solid android:color="#7D65FC"/>
</shape>
```

【文件 2-11】res/drawable/circle_white.xml

```xml
<?xml version="1.0" encoding="utf-8"?>
<shape xmlns:android="http://schemas.android.com/apk/res/android" android:shape="oval">
    <solid android:color="#FFFFFF"/>
</shape>
```

2.4.2 向导界面（一）

设置向导第一个界面使用了两个线性布局，将界面分为两大部分，用于控制布局所占的权重。上部分的线性布局嵌套了一个相对布局和一个线性布局，其中相对布局中放一个安全图标和文字"手机防盗向导"；线性布局中又嵌套了两个线性布局，分别用于放置 SIM 卡变更报警、GPS 追踪的提示文字及图标，以及远程锁屏、远程删除数据的提示文字及图标；下半部分的线性布局中通过<include/>引入了小圆点布局。设置向导一的图形化界面如图 2-9 所示。

图 2-9 所示设置向导一对应的布局文件如【文件 2-12】所示。

图 2-9 设置向导一

【文件 2-12】activity_setup1.xml

```xml
<?xml version="1.0" encoding="utf-8"?>
<LinearLayout xmlns:android="http://schemas.android. com/apk/res/android"
    android:layout_width="match_parent"
    android:layout_height="match_parent"
    android:orientation="vertical" >
    <LinearLayout
```

```xml
        android:layout_width="match_parent"
        android:layout_height="match_parent"
        android:layout_weight="100"
        android:orientation="vertical" >
    <RelativeLayout
        android:layout_width="match_parent"
        android:layout_height="160dp"
        android:background="@color/purple" >
        <ImageView
            android:id="@+id/imgv"
            android:layout_width="80dp"
            android:layout_height="80dp"
            android:layout_centerInParent="true"
            android:background="@drawable/security_phone_setup1" />
        <TextView
            android:layout_width="match_parent"
            android:layout_height="wrap_content"
            android:layout_below="@+id/imgv"
            android:layout_marginTop="10dp"
            android:gravity="center"
            android:text="手机防盗向导"
            android:textColor="@android:color/white"
            android:textSize="18sp" />
    </RelativeLayout>
    <LinearLayout
        android:layout_width="match_parent"
        android:layout_height="match_parent"
        android:background="@android:color/white"
        android:orientation="vertical" >
        <LinearLayout
            android:layout_width="match_parent"
            android:layout_height="0dp"
            android:layout_gravity="center"
            android:layout_weight="1"
            android:gravity="center_vertical"
            android:orientation="horizontal" >
            <TextView
                style="@style/textview16sp"
```

```xml
            android:layout_width="0dp"
            android:layout_weight="1"
            android:drawablePadding="5dp"
            android:drawableTop="@drawable/sim_alarm_icon"
            android:gravity="center"
            android:text="SIM卡变更报警"
            android:textColor="@color/purple" />
        <TextView
            style="@style/textview16sp"
            android:layout_width="0dp"
            android:layout_weight="1"
            android:drawablePadding="5dp"
            android:drawableTop="@drawable/location_icon"
            android:gravity="center"
            android:text="GPS追踪"
            android:textColor="@color/purple" />
    </LinearLayout>
    <LinearLayout
        android:layout_width="match_parent"
        android:layout_height="0dp"
        android:layout_gravity="center"
        android:layout_weight="1"
        android:gravity="center_vertical"
        android:orientation="horizontal" >
        <TextView
            style="@style/textview16sp"
            android:layout_width="0dp"
            android:layout_weight="1"
            android:drawablePadding="5dp"
            android:drawableTop="@drawable/lock_screen_icon"
            android:gravity="center"
            android:text="远程锁屏"
            android:textColor="@color/purple" />
        <TextView
            style="@style/textview16sp"
            android:layout_width="0dp"
            android:layout_weight="1"
            android:drawablePadding="5dp"
```

```xml
                android:drawableTop="@drawable/delete_data"
                android:gravity="center"
                android:text="远程删除数据"
                android:textColor="@color/purple" />
        </LinearLayout>
    </LinearLayout>
</LinearLayout>
<LinearLayout
    android:layout_width="match_parent"
    android:layout_height="55dp"
    android:layout_weight="10"
    android:background="@color/purple"
    android:gravity="center" >
    <include layout="@layout/setup_radiogroup" />
</LinearLayout>
</LinearLayout>
```

上述布局中，使用了很多 TextView 控件，为了减少代码的重复量，将其样式定义在 styles.xml 文件中，具体代码如【文件 2-13】所示。

【文件 2-13】res/values/styles.xml

```xml
<resources>
    <!-- 高宽都包裹内容 -->
    <style name="wrapcontent">
        <item name="android:layout_width">wrap_content</item>
        <item name="android:layout_height">wrap_content</item>
    </style>
    <!-- TextView 16sp 样式 -->
    <style name="textview16sp" parent="wrapcontent">
        <item name="android:textSize">16sp</item>
        <item name="android:gravity">center_vertical</item>
    </style>
    <!-- TextView 14sp 样式 -->
    <style name="textview14sp" parent="wrapcontent">
        <item name="android:textSize">14sp</item>
        <item name="android:gravity">center_vertical</item>
    </style>
</resources>
```

需要注意的是，textview14sp 样式暂时不会用到，它是在获取联系人的 Item 布局中使用的，暂且一起进行定义。

2.4.3 向导界面（二）

设置向导第二个界面同样分为两大部分，上部分的线性布局包含一个相对布局、两个线性布局，其中相对布局用于放置 SIM 卡图标及文字；第一个线性布局放置一个图标及文字提示信息；第二个线性布局放置一个 Button 按钮，用于点击绑定 SIM 卡。下部分的线性布局中引入了小圆点布局。设置向导二的图形化界面如图 2-10 所示。

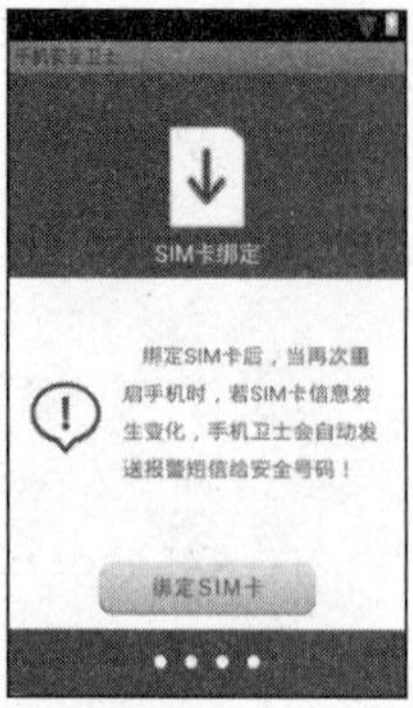

图 2-10 所示设置向导二对应的布局文件如【文件 2-14】所示。

【文件 2-14】activity_setup2.xml

图 2-10 设置向导二

```xml
<?xml version="1.0" encoding="utf-8"?>
<LinearLayout xmlns:android="http://schemas.android.com/apk/res/android"
    android:layout_width="match_parent"
    android:layout_height="match_parent"
    android:orientation="vertical" >
    <LinearLayout
        android:layout_width="match_parent"
        android:layout_height="match_parent"
        android:layout_weight="100"
        android:background="@android:color/white"
        android:orientation="vertical" >
        <RelativeLayout
            android:layout_width="match_parent"
            android:layout_height="160dp"
            android:background="@color/purple" >
            <ImageView
                android:id="@+id/imgv"
                android:layout_width="80dp"
                android:layout_height="80dp"
                android:layout_centerInParent="true"
                android:background="@drawable/sim_setup" />
            <TextView
                android:layout_width="match_parent"
                android:layout_height="wrap_content"
                android:layout_below="@+id/imgv"
                android:layout_marginTop="10dp"
                android:gravity="center"
                android:text="SIM卡绑定"
                android:textColor="@android:color/white"
                android:textSize="18sp" />
```

```xml
    </RelativeLayout>
    <LinearLayout
        android:layout_width="match_parent"
        android:layout_height="0dp"
        android:layout_gravity="center_vertical"
        android:layout_weight="4"
        android:gravity="center_vertical"
        android:orientation="horizontal" >
        <ImageView
            android:layout_width="60dp"
            android:layout_height="60dp"
            android:layout_gravity="center_vertical|left"
            android:layout_marginLeft="15dp"
            android:background="@drawable/recomand_icon" />
        <TextView
            style="@style/textview16sp"
            android:layout_marginLeft="15dp"
            android:layout_marginRight="15dp"
            android:lineSpacingMultiplier="1.5"
            android:text="@string/_sim_sim_bind_info"
            android:textColor="@color/purple"
            android:textScaleX="1.1" />
    </LinearLayout>
    <LinearLayout
        android:layout_width="match_parent"
        android:layout_height="0dp"
        android:layout_gravity="center_vertical"
        android:layout_weight="1"
        android:gravity="center" >
        <Button
            android:id="@+id/btn_bind_sim"
            android:layout_width="180dp"
            android:layout_height="45dp"
            android:layout_gravity="center_horizontal"
            android:layout_marginBottom="20dp"
            android:background="@drawable/sim_bind_selector"
            android:gravity="center" />
    </LinearLayout>
```

```xml
    </LinearLayout>
    <LinearLayout
        android:layout_width="match_parent"
        android:layout_height="55dp"
        android:layout_weight="10"
        android:background="@color/purple"
        android:gravity="center" >
        <include layout="@layout/setup_radiogroup" />
    </LinearLayout>
</LinearLayout>
```

上述代码中,绑定 SIM 卡的提示文字是定义在 strings.xml 文件中的,通过属性 android:text="@string/_sim_sim_bind_info"将其引入,具体代码如【文件 2-15】所示。

【文件 2-15】res/values/strings.xml

```xml
<?xml version="1.0" encoding="utf-8"?>
<resources>
    <string name="app_name">手机安全卫士</string>
    <string name="hello_world">Hello world!</string>
    <string name="action_settings">Settings</string>
    <string name="_sim_sim_bind_info">"绑定SIM卡后,当再次重启手机时,若SIM卡信息发生变化,手机卫士会自动发送报警短信给安全号码! "</string>
</resources>
```

当前布局下方"绑定 SIM 卡"按钮使用了背景选择器,在按钮按下与释放时显示不同颜色的图片,背景选择器的代码如【文件 2-16】所示。

【文件 2-16】res/drawable/sim_bind_selector

```xml
<?xml version="1.0" encoding="utf-8"?>
<selector xmlns:android="http://schemas.android.com/apk/res/android">
    <item android:drawable="@drawable/sim_bind_p" android:state_pressed="true"/>
    <item android:drawable="@drawable/sim_bind_e" android:state_enabled="false"/>
    <item android:drawable="@drawable/sim_bind_n"/>
</selector>
```

从上述代码可以看出,默认情况下按钮显示正常背景(sim_bind_n.png),当按钮按下时,使用蓝色的背景图片(sim_bind_p.png),当按钮弹起时,使用灰色的背景图片(sim_bind_e.png)。

值得一提的是,后面代码中很多按钮都使用了背景选择器,用于增强用户体验,在按钮按下与弹起时背景进行变化。如果在代码中动态设置,相对比较麻烦,因此通常使用 Android 提供的 Selector 选择器。

2.4.4 向导界面（三）

设置向导第三个界面也分为两大部分，上部分的线性布局包含一个相对布局和一个线性布局，其中相对布局中放了选择联系人图标以及文字，线性布局中放置了一个文本编辑框以及一个 Button 按钮。下部分的线性布局中引入了小圆点布局。设置向导三的图形化界面如图 2-11 所示。

图 2-11 所示设置向导三对应的布局文件如【文件 2-17】所示。

【文件 2-17】activity_setup3.xml

图 2-11 设置向导三

```xml
<?xml version="1.0" encoding="utf-8"?>
<LinearLayout xmlns:android="http://schemas.android.com/apk/res/android"
    android:layout_width="match_parent"
    android:layout_height="match_parent"
    android:orientation="vertical" >
    <LinearLayout
        android:layout_width="match_parent"
        android:layout_height="match_parent"
        android:layout_weight="100"
        android:background="@android:color/white"
        android:orientation="vertical" >
        <RelativeLayout
            android:layout_width="match_parent"
            android:layout_height="160dp"
            android:background="@color/purple" >
            <ImageView
                android:id="@+id/imgv"
                android:layout_width="80dp"
                android:layout_height="80dp"
                android:layout_centerInParent="true"
                android:background="@drawable/set_safephone_icon"/>
            <TextView
                android:layout_width="match_parent"
                android:layout_height="wrap_content"
                android:layout_below="@+id/imgv"
                android:layout_marginTop="10dp"
                android:gravity="center"
                android:text="选择安全联系人"
                android:textColor="@android:color/white"
```

```xml
                android:textSize="18sp" />
        </RelativeLayout>
        <LinearLayout
            android:layout_width="match_parent"
            android:layout_height="55dp"
            android:layout_margin="15dp"
            android:gravity="center_vertical"
            android:orientation="horizontal" >
            <EditText
                android:id="@+id/et_inputphone"
                android:layout_width="match_parent"
                android:layout_height="50dp"
                android:layout_weight="5"
                android:background="@drawable/coner_white_rec"
                android:drawableLeft="@drawable/contact_et_left_icon"
                android:hint=" 请输入安全号码"
                android:inputType="phone"
                android:textColorHint="@color/purple"/>
            <Button
                android:id="@+id/btn_addcontact"
                android:layout_width="55dp"
                android:layout_height="45dp"
                android:layout_marginLeft="10dp"
                android:layout_weight="1"
                android:background="@drawable/add"/>
        </LinearLayout>
    </LinearLayout>
    <LinearLayout
        android:layout_width="match_parent"
        android:layout_height="55dp"
        android:layout_weight="10"
        android:background="@color/purple"
        android:gravity="center" >
        <include layout="@layout/setup_radiogroup"/>
    </LinearLayout>
</LinearLayout>
```

2.4.5 向导界面（四）

设置向导第四个界面分为两大部分，上部分的线性布局包含两个相对布局，其中第一个相对布局中放置了设置完成图标及提示文字；第二个相对布局中放置一个安全图标及一个 ToggleButton 按钮，用于开启和关闭防盗保护。下部分的线性布局中引入了小圆点布局。设置向导四的图形化界面如图 2-12 所示。

图 2-12 所示设置向导四对应的布局文件如【文件 2-18】所示。

图 2-12 设置向导四

【文件 2-18】activity_setup4.xml

```xml
<?xml version="1.0" encoding="utf-8"?>
<LinearLayout xmlns:android="http://schemas.android.com/apk/res/android"
    android:layout_width="match_parent"
    android:layout_height="match_parent"
    android:orientation="vertical" >
    <LinearLayout
        android:layout_width="match_parent"
        android:layout_height="match_parent"
        android:layout_weight="100"
        android:orientation="vertical" >
        <RelativeLayout
            android:layout_width="match_parent"
            android:layout_height="0dp"
            android:layout_weight="4"
            android:background="@color/purple" >
            <ImageView
                android:id="@+id/imgv"
                android:layout_width="80dp"
                android:layout_height="80dp"
                android:layout_centerInParent="true"
                android:background="@drawable/setup_setting_complete"/>
            <TextView
                android:layout_width="match_parent"
                android:layout_height="wrap_content"
                android:layout_below="@+id/imgv"
                android:layout_marginTop="20dp"
                android:gravity="center"
                android:text="恭喜！设置完成"
                android:textColor="@android:color/white"
                android:textScaleX="1.1"
```

```xml
            android:textSize="18sp"/>
        </RelativeLayout>
        <RelativeLayout
            android:layout_width="match_parent"
            android:layout_height="0dp"
            android:layout_weight="1"
            android:background="@android:color/white">
            <TextView
                android:id="@+id/tv_setup4_status"
                style="@style/textview16sp"
                android:layout_centerVertical="true"
                android:layout_marginLeft="10dp"
                android:drawableLeft="@drawable/security_phone"
                android:drawablePadding="10dp"
                android:textColor="@color/purple"/>
            <ToggleButton
                android:id="@+id/togglebtn_securityfunction"
                android:layout_width="70dp"
                android:layout_height="30dp"
                android:layout_alignParentRight="true"
                android:layout_centerVertical="true"
                android:layout_marginRight="27dp"
                android:background="@drawable/toggle_btn_bg_selector"
                android:textOff=""
                android:textOn=""/>
        </RelativeLayout>
    </LinearLayout>
    <LinearLayout
        android:layout_width="match_parent"
        android:layout_height="55dp"
        android:layout_weight="10"
        android:background="@color/purple"
        android:gravity="center" >
        <include layout="@layout/setup_radiogroup"/>
    </LinearLayout>
</LinearLayout>
```

上述布局文件中，使用了一个 ToggleButton 控件，该控件用于展示一个可切换的按钮，控制条目的开启与关闭，android:textOn 表示选中时按钮的文本，android:textOff 属性用于表示未选中时按钮的文本，这两个文本都在代码中进行控制。并且还通过 android:background 属性为按钮设置背景图片，在指定背景图片时，是通过图片选择器控制的，具体代码如【文件 2-19】所示。

【文件 2-19】res/drawable/toggle_btn_bg_selector.xml

```xml
<?xml version="1.0" encoding="utf-8"?>
<selector xmlns:android="http://schemas.android.com/apk/res/android" >
  <item android:state_checked="true"
                  android:drawable="@drawable/swtich_btn_on"/>
  <item  android:state_checked="false"
                  android:drawable="@drawable/switch_btn_off"/>
</selector>
```

上述选择器中，当按钮处于被选中状态时，为按钮指定一张紫色的背景图片（swtich_btn_on.png），当按钮未被选中时，为按钮指定一张灰色的背景图（swtich_btn_off.png）。这样就让人觉得当点击按钮时，按钮在不停地切换。

2.4.6 指令界面

设置向导的防盗指令界面由三个相对布局以及一个线性布局组成，三个相对布局分别用于显示安全号码、防盗保护是否开启和重新进入设置向导。下面的线性布局嵌套了四个相对布局，分别用于显示播放报警音乐指令、GPS 追踪指令、远程锁屏指令和远程删除数据指令。设置向导指令的图形化界面如图 2-13 所示。

图 2-13 所示防盗指令对应的布局文件如【文件 2-20】所示。

【文件 2-20】activity_lostfind.xml

图 2-13 防盗指令界面

```xml
<?xml version="1.0" encoding="utf-8"?>
<LinearLayout xmlns:android="http://schemas.android.com/apk/res/android"
  android:layout_width="match_parent"
  android:layout_height="match_parent"
  android:background="@android:color/white"
  android:orientation="vertical">
  <include layout="@layout/titlebar"/>
  <RelativeLayout
     android:layout_width="match_parent"
     android:layout_height="50dp" >
     <TextView
         style="@style/textview16sp"
         android:layout_alignParentLeft="true"
         android:layout_centerVertical="true"
         android:layout_marginLeft="10dp"
         android:text="安全号码"
```

```xml
        android:textColor="@color/purple"/>
    <TextView
        android:id="@+id/tv_safephone"
        style="@style/textview16sp"
        android:layout_alignParentRight="true"
        android:layout_centerVertical="true"
        android:layout_marginRight="15dp"
        android:textColor="@color/purple"/>
</RelativeLayout>
<View
    android:layout_width="match_parent"
    android:layout_height="1.0px"
    android:background="#10000000"/>
<RelativeLayout
    android:layout_width="match_parent"
    android:layout_height="50dp">
    <TextView
        android:id="@+id/tv_lostfind_protectstauts"
        style="@style/textview16sp"
        android:layout_alignParentLeft="true"
        android:layout_centerVertical="true"
        android:layout_marginLeft="10dp"
        android:text="防盗保护是否开启"
        android:textColor="@color/purple"/>
    <ToggleButton
        android:id="@+id/togglebtn_lostfind"
        android:layout_width="70dp"
        android:layout_height="30dp"
        android:layout_alignParentRight="true"
        android:layout_centerVertical="true"
        android:layout_marginRight="10dp"
        android:background="@drawable/toggle_btn_bg_selector"
        android:textOff=""
        android:textOn=""/>
</RelativeLayout>
<View
    android:layout_width="match_parent"
    android:layout_height="1.0px"
    android:background="#10000000"/>
<RelativeLayout
```

```xml
        android:id="@+id/rl_inter_setup_wizard"
        android:layout_width="match_parent"
        android:layout_height="50dp" >
        <TextView
            style="@style/textview16sp"
            android:layout_alignParentLeft="true"
            android:layout_centerVertical="true"
            android:layout_marginLeft="10dp"
            android:text="重新进入设置向导"
            android:textColor="@color/purple"/>
        <ImageView
            android:layout_width="wrap_content"
            android:layout_height="wrap_content"
            android:layout_alignParentRight="true"
            android:layout_centerVertical="true"
            android:layout_marginRight="10dp"
            android:background="@drawable/arrow_right"/>
</RelativeLayout>
<View
    android:layout_width="match_parent"
    android:layout_height="1.0px"
    android:background="#10000000"/>
<LinearLayout
    android:layout_width="wrap_content"
    android:layout_height="wrap_content"
    android:layout_marginLeft="10dp"
    android:layout_marginRight="10dp"
    android:layout_marginTop="30dp"
    android:orientation="vertical">
    <TextView
        android:layout_width="match_parent"
        android:layout_height="35dp"
        android:background="@drawable/round_purple_tv_bg"
        android:gravity="center"
        android:text="短信指令功能简介"
        android:textColor="@android:color/white"
        android:textSize="16sp"/>
    <RelativeLayout
        android:layout_width="match_parent"
        android:layout_height="wrap_content"
```

```xml
            android:layout_marginTop="15dp" >
            <TextView
                style="@style/textview16sp"
                android:layout_alignParentRight="true"
                android:layout_marginRight="10dp"
                android:text="#*alarm*#"
                android:textColor="@color/purple"/>
            <TextView
                style="@style/textview16sp"
                android:layout_alignParentLeft="true"
                android:layout_centerVertical="true"
                android:layout_marginLeft="10dp"
                android:drawableLeft="@drawable/sim_alarm_icon_small"
                android:drawablePadding="5dp"
                android:text="播放报警音乐"
                android:textColor="@color/purple"/>
        </RelativeLayout>
        <RelativeLayout
            android:layout_width="match_parent"
            android:layout_height="wrap_content"
            android:layout_marginTop="15dp">
            <TextView
                style="@style/textview16sp"
                android:layout_alignParentRight="true"
                android:layout_marginRight="10dp"
                android:text="#*location*#"
                android:textColor="@color/purple"/>
            <TextView
                style="@style/textview16sp"
                android:layout_alignParentLeft="true"
                android:layout_centerVertical="true"
                android:layout_marginLeft="10dp"
                android:drawableLeft="@drawable/location_icon_small"
                android:drawablePadding="5dp"
                android:text="GPS 追踪"
                android:textColor="@color/purple"/>
        </RelativeLayout>
        <RelativeLayout
            android:layout_width="match_parent"
            android:layout_height="wrap_content"
```

```xml
            android:layout_marginTop="15dp">
            <TextView
                style="@style/textview16sp"
                android:layout_alignParentRight="true"
                android:layout_marginRight="10dp"
                android:text="#*lockScreen*#"
                android:textColor="@color/purple"/>
            <TextView
                style="@style/textview16sp"
                android:layout_alignParentLeft="true"
                android:layout_centerVertical="true"
                android:layout_marginLeft="10dp"
                android:drawableLeft="@drawable/lock_screen_icon_small"
                android:drawablePadding="5dp"
                android:text="远程锁屏"
                android:textColor="@color/purple"/>
        </RelativeLayout>
        <RelativeLayout
            android:layout_width="match_parent"
            android:layout_height="wrap_content"
            android:layout_marginTop="15dp">
            <TextView
                style="@style/textview16sp"
                android:layout_alignParentRight="true"
                android:layout_marginRight="10dp"
                android:text="#*wipedata*#"
                android:textColor="@color/purple"/>
            <TextView
                style="@style/textview16sp"
                android:layout_alignParentLeft="true"
                android:layout_centerVertical="true"
                android:layout_marginLeft="10dp"
                android:drawableLeft="@drawable/delete_data_small"
                android:drawablePadding="5dp"
                android:text="远程删除数据"
                android:textColor="@color/purple"/>
        </RelativeLayout>
    </LinearLayout>
</LinearLayout>
```

上述布局文件中，使用 include 属性引入了一个 titlebar.xml 布局文件，该布局文件用于设置标题栏，在其他界面中也可以使用。标题栏的代码如【文件2-21】所示。

【文件2-21】titlebar.xml

```xml
<?xml version="1.0" encoding="utf-8"?>
<RelativeLayout xmlns:android="http://schemas.android.com/apk/res/android"
    android:layout_width="match_parent"
    android:layout_height="55dp"
    android:background="@color/purple">
    <TextView
        android:id="@+id/tv_title"
        android:layout_width="wrap_content"
        android:layout_height="wrap_content"
        android:textSize="20sp"
        android:textColor="@android:color/white"
        android:layout_centerInParent="true"/>
    <ImageView
        android:id="@+id/imgv_leftbtn"
        android:layout_width="35dp"
        android:layout_height="35dp"
        android:layout_alignParentLeft="true"
        android:layout_centerVertical="true"
        android:layout_marginLeft="10dp"/>
    <ImageView
        android:id="@+id/imgv_rightbtn"
        android:layout_width="wrap_content"
        android:layout_height="wrap_content"
        android:layout_alignParentRight="true"
        android:layout_centerVertical="true"
        android:layout_marginRight="10dp"/>
</RelativeLayout>
```

上述布局文件中，放置了一个TextView控件以及两个ImageView控件，其中TextView控件用于展示标题，ImageView控件用于在标题文字左侧或右侧放置返回按钮。

在防盗指令界面中，用于显示"短信指令功能简介"的TextView控件通过android:background属性引入自定义属性，用于指定TextView控件的背景形状，具体代码如【文件2-22】所示。

【文件2-22】res/drawable/round_purple_tv_bg.xml

```xml
<?xml version="1.0" encoding="utf-8"?>
<shape xmlns:android="http://schemas.android.com/apk/res/android"
    android:shape="rectangle"
```

```
    <solid android:color="@color/purple"/>
    <corners android:topLeftRadius="5dp"
        android:topRightRadius="5dp"
        android:bottomLeftRadius="0dp"
        android:bottomRightRadius="0dp"/>
</shape>
```

在上述选择器中，通过<solid>标签指定当前填充颜色为紫色，<corners>标签用于指定圆角样式，其中android:topLeftRadius属性用于设置左上圆角的半径，android:topRightRadius属性用于设置右上圆角的半径，android:bottomLeftRadius 属性用于设置左下圆角的半径，android:bottomRightRadius属性用于设置右下圆角的半径。

2.5 设置向导功能

设置向导主要用于展示手机防盗模块，以及绑定 SIM 卡、设置安全号码、开启防盗保护功能等，本节将针对设置向导功能进行详细讲解。

2.5.1 滑屏动画

在 Android 系统中，通过手势识别切换界面时，通常会在界面切换时加入动画，以提高用户的体验效果，这种动画一般都采用平移动画，当下一个界面进入时，上一个界面移出屏幕，如图 2-14 所示。

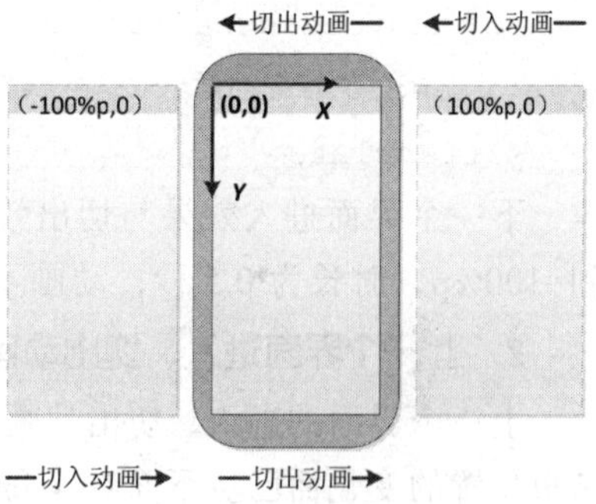

从图 2-14 可以看出，屏幕左上角的坐标为（0,0），进入屏幕的界面坐标为（100%p,0），从屏幕切出界面的坐标为（-100%p,0）。需要注意的是，p 是指屏幕，100%p 表示整个屏幕，切入界面和切出界面都是以整个屏幕为单位计算的。

界面切换的平移动画有四个，分别是下一个界面进入与切出效果，以及上一个界面进入与切出效果，接下来分别定义这四个动画文件。

图 2-14 动画效果图

1. 下一个界面进入、切出动画

在 res 目录中添加一个 anim 文件夹，在该文件夹中定义下一个界面进入的动画效果(next_in.xml)，具体代码如【文件 2-23】所示。

【文件 2-23】res/anim/next_in.xml

```
<?xml version="1.0" encoding="utf-8"?>
<!-- 显示下一步，下一个页面进来的效果 -->
<translate xmlns:android="http://schemas.android.com/apk/res/android"
    android:fromXDelta="100%p"
    android:toXDelta="0"
    android:fromYDelta="0"
```

```xml
    android:toYDelta="0"
    android:duration="500"
    android:repeatCount="0">
</translate>
```

上述代码中，android:fromXDelta="100%p"，android:toXDelta="0"属性表示当前界面从 X 轴坐标 100%p 移动到 0，android:duration="500"表示动画执行时长为 500 毫秒，repeatCount="0" 表示动画不重复执行。

在 anim 文件夹下创建下一个界面切出的动画效果（next_out.xml），具体代码如【文件 2-24】所示。

【文件 2-24】res/anim/next_out.xml

```xml
<?xml version="1.0" encoding="utf-8"?>
<!-- 显示下一步，当前页面出去的效果 -->
<translate xmlns:android="http://schemas.android.com/apk/res/android"
    android:fromXDelta="0"
    android:toXDelta="-100%p"
    android:fromYDelta="0"
    android:toYDelta="0"
    android:duration="500"
    android:repeatCount="0">
</translate>
```

下一个界面进入效果与切出效果代码类似，只不过当前界面是从 X 轴坐标 0 位置移动到 -100%p，时长为 0.5 秒，动画不重复执行。

2. 上一个界面进入、切出动画

上一个界面的进入、切出动画与上面的代码非常类似，只不过是 X 轴坐标的起始位置与结束位置的变化而已。下面在 anim 文件夹下创建上一个界面进入的动画效果（pre_in.xml），具体代码如【文件 2-25】所示。

【文件 2-25】res/anim/pre_in.xml

```xml
<?xml version="1.0" encoding="utf-8"?>
<!-- 显示上一步，上一个页面进来的效果 -->
<translate xmlns:android="http://schemas.android.com/apk/res/android"
    android:fromXDelta="-100%p"
    android:toXDelta="0"
    android:fromYDelta="0"
    android:toYDelta="0"
    android:duration="500"
    android:repeatCount="0">
</translate>
```

在 anim 文件夹下创建上一个界面切出的动画效果（pre_in.xml），具体代码如【文件 2-26】

所示。

【文件 2-26】res/anim/pre_out.xml

```
<?xml version="1.0" encoding="utf-8"?>
<!-- 显示上一步,当前页面出去的效果 -->
<translate xmlns:android="http://schemas.android.com/apk/res/android"
    android:fromXDelta="0"
    android:toXDelta="100%p"
    android:fromYDelta="0"
    android:toYDelta="0"
    android:duration="500"
    android:repeatCount="0" >
</translate>
```

至此,所有的动画效果已经创建完成,只需在切换屏幕时引入即可。

2.5.2 手势滑动

设置向导中最核心的内容就是通过手势识别切换界面,当手指落在手机屏幕上按住屏幕快速滑动屏幕即可切换。为了更生动形象地展示切换效果,通过一个图例进行展示,具体如图 2-15 所示。

从图 2-15 可以看出,当手指在左侧按下向右快速滑动时,上一屏进入当前界面,当手指在右侧按下快速向左侧滑动时,下一屏进入当前界面。需要注意的是,在滑动过程中需要过滤一些无效的动作(如滑动太慢)。

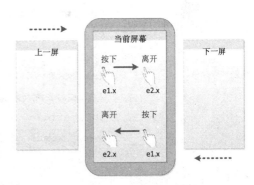

图 2-15 屏幕切换效果图

从功能介绍中可知,设置向导的每一个界面都需要进行滑动,也就是说每个 Activity 都要有手势识别器。为了避免代码的重复编写,可以定义一个父类实现手势识别功能,其他类继承该类即可。设置向导界面的父类如【文件 2-27】所示。

【文件 2-27】BaseSetUpActivity.java

```
1  public abstract class BaseSetUpActivity extends Activity{
2      public SharedPreferences sp;
3      private GestureDetector mGestureDetector;
4      @Override
5      protected void onCreate(Bundle savedInstanceState){
6          super.onCreate(savedInstanceState);
7          requestWindowFeature(Window.FEATURE_NO_TITLE);
8          sp = getSharedPreferences("config", MODE_PRIVATE);
9          //1.初始化手势识别器
```

```java
10      mGestureDetector = new GestureDetector(this,
11        new GestureDetector.SimpleOnGestureListener(){
12          //e1 代表手指第一次触摸屏幕的事件，e2 代表手指离开屏幕一瞬间的事件
13          // velocityX 水平方向的速度 单位 pix/s，velocityY 竖直方向的速度
14          @Override
15          public boolean onFling(MotionEvent e1, MotionEvent e2,float
16          velocityX,float velocityY){
17              if(Math.abs(velocityX)<200){
18                  Toast.makeText(getApplicationContext(),"无效动作,移动太慢",
19                  0).show();
20                  return true;
21              }
22              if((e2.getRawX()-e1.getRawX())>200){
23                  //从左向右滑动屏幕，显示上一个界面
24                  showPre();
25                  overridePendingTransition(R.anim.pre_in,R.anim.pre_out);
26                  return true;
27              }
28              if((e1.getRawX()-e2.getRawX())>200){
29                  //从右向左滑动屏幕，显示下一个界面
30                  showNext();
31                  overridePendingTransition(R.anim.next_in,R.anim.next_out);
32                  return true;
33              }
34          return super.onFling(e1,e2,velocityX,velocityY);
35          }
36      });
37  }
38  public abstract void showNext();
39  public abstract void showPre();
40  //2.用手势识别器去识别事件
41  @Override
42  public boolean onTouchEvent(MotionEvent event){
43      //分析手势事件
44      mGestureDetector.onTouchEvent(event);
45      return super.onTouchEvent(event);
46  }
47  /**
48   * 开启新的activity并且关闭自己
```

```
49      * @param cls 新的activity的字节码
50      */
51     public void startActivityAndFinishSelf(Class<?>cls){
52         Intent intent=new Intent(this,cls);
53         startActivity(intent);
54         finish();
55     }
56 }
```

代码说明：

- 第 10～37 行代码用于初始化手势识别器，并实现 onFling()方法，在该方法中通过 if 语句对手势的滑动效果进行判断，如果 if(Math.abs(velocityX)<200)时表示 1 秒移动小于 200 像素，移动速度太慢界面不切换；如果 if((e2.getRawX()-e1.getRawX())>200)时，则从左向右滑动屏幕显示上一个界面并显示动画效果；如果 if((e1.getRawX()-e2.getRawX())> 200) 时，则从右向左滑动屏幕，显示下一个界面并显示动画效果。
- 第 38～39 行代码定义了一个 showNext()方法和一个 showPre()方法，分别用于展示下一个页面和上一个页面。这两个方法都是抽象的，需要子类重写。
- 第 42～46 行重写了 onTouchEvent()方法，并通过手势识别器分析屏幕上的手势事件。
- 第 51～55 行的 startActivityAndFinishSelf(Class<?> cls)方法用于在当前界面开启一个新的 Activity 时,把当前的 Activity 关闭,其中传入的参数 Class<?> cls 可以是一个 Activity。

2.5.3 向导功能（一）

当父类创建好之后接下来实现设置向导第一个界面的逻辑，由于第一个界面只是单纯的用于展示，因此逻辑较为简单，具体代码如【文件 2-28】所示。

【文件 2-28】SetUp1Activity.java

```
1  public class SetUp1Activity extends BaseSetUpActivity{
2      @Override
3      protected void onCreate(Bundle savedInstanceState){
4          super.onCreate(savedInstanceState);
5          setContentView(R.layout.activity_setup1);
6          initView();
7      }
8      private void initView(){
9          // 设置第一个小圆点的颜色
10         ((RadioButton) findViewById(R.id.rb_first)).setChecked(true);
11     }
12     @Override
13     public void showNext(){
14         startActivityAndFinishSelf(SetUp2Activity.class);
15     }
```

```
16     @Override
17     public void showPre(){
18         Toast.makeText(this,"当前页面已经是第一页",0).show();
19     }
20 }
```

代码说明：

- 第8~11行的initView()方法用于初始化控件，将界面小圆点的状态置为选中状态，此时小圆点的颜色为紫色。
- 第13~15行的showNext()方法重写了BaseSetUpActivity中的showNext()方法，用于开启SetUp2Activity并关闭当前Activity。
- 第17~19行 showPre()方法重写了BaseSetUpActivity中的showPre()方法，当手指在屏幕上从左向右滑动时用于弹出提示信息，说明当前页面是第一页不能显示前一页。

2.5.4 向导功能（二）

设置向导第二个界面主要用于绑定SIM卡，相比第一个界面来说逻辑较为复杂，具体代码如【文件2-29】所示。

【文件2-29】SetUp2Activity.java

```
1  public class SetUp2Activity extends BaseSetUpActivity implements
2  OnClickListener{
3      private TelephonyManager mTelephonyManager;
4      private Button mBindSIMBtn;
5      @Override
6      protected void onCreate(Bundle savedInstanceState){
7          super.onCreate(savedInstanceState);
8          setContentView(R.layout.activity_setup2);
8          mTelephonyManager=(TelephonyManager) getSystemService(TELEPHONY_SERVICE);
10         initView();
11     }
12     private void initView(){
13         //设置第二个小圆点的颜色
14         ((RadioButton) findViewById(R.id.rb_second)).setChecked(true);
15         mBindSIMBtn=(Button) findViewById(R.id.btn_bind_sim);
16         mBindSIMBtn.setOnClickListener(this);
17         if(isBind()){
18             mBindSIMBtn.setEnabled(false);
19         }else{
20             mBindSIMBtn.setEnabled(true);
21         }
22     }
```

```java
23  private boolean isBind(){
24      String simString=sp.getString("sim", null);
25      if(TextUtils.isEmpty(simString)){
26          return false;
27      }
28      return true;
29  }
30  @Override
31  public void showNext(){
32      if(!isBind()){
33          Toast.makeText(this,"您还没有绑定SIM卡!",0).show();
34          return;
35      }
36      startActivityAndFinishSelf(SetUp3Activity.class);
37  }
38  @Override
39  public void showPre(){
40      startActivityAndFinishSelf(SetUp1Activity.class);
41  }
42  @Override
43  public void onClick(View v){
44      switch(v.getId()){
45      case R.id.btn_bind_sim:
46          //绑定SIM卡
47          bindSIM();
48          break;
49      }
50  }
51  /**
52   * 绑定sim卡
53   */
54  private void bindSIM(){
55      if(!isBind()){
56          String simSerialNumber=mTelephonyManager.getSimSerialNumber();
57          Editor edit=sp.edit();
58          edit.putString("sim",simSerialNumber);
59          edit.commit();
60          Toast.makeText(this,"SIM卡绑定成功!",0).show();
61          mBindSIMBtn.setEnabled(false);
```

```
62          }else{
63              //已经绑定，提醒用户
64              Toast.makeText(this, "SIM卡已经绑定! ",0).show();
65              mBindSIMBtn.setEnabled(false);
66          }
67      }
68 }
```

代码说明：

- 第 12~22 行的 initView()方法用于初始化控件，指定第 2 个小圆点被选中，并判断 SIM 卡是否绑定，如果已绑定则将该 Button 按钮设置为不可用。
- 第 54~67 行的 bindSIM()方法用于绑定 SIM 卡，通过 if 语句判断如果当前 SIM 卡没有绑定，则获取手机的 SIM 卡串号，并存入 SharedPreferences 对象中，如果绑定了则弹出 Toast 进行，并将按钮设置为不可用。

在检测 SIM 卡是否发生变化时，可以使用 Application 类。该类是 Android 框架的一个系统组件，当 Android 程序启动时系统会创建 Application 对象（单例模式的一个类，需要在 application 标签增加 name 属性，并添加 Application 名字即可）。正是由于它的这种特性，可以将检测 SIM 卡是否变更的方法放在 Application 的 onCreate()方法中，当程序启动时就会检测 SIM 卡是否变更。接下来创建一个 Application 子类 App，具体代码如【文件 2-30】所示。

【文件 2-30】App.java

```
1  public class App extends Application{
2      @Override
3      public void onCreate(){
4          super.onCreate();
5          correctSIM();
6      }
7      public void correctSIM(){
8          //检查SIM卡是否发生变化
9          SharedPreferences sp=getSharedPreferences("config",Context.
10         MODE_PRIVATE);
11         //获取防盗保护的状态
12         boolean protecting=sp.getBoolean("protecting",true);
13         if(protecting){
14             //得到绑定的sim卡串号
15             String bindsim=sp.getString("sim","");
16             //得到手机现在的sim卡串号
17             TelephonyManager tm=(TelephonyManager)
18             getSystemService(Context.TELEPHONY_SERVICE);
19             //为了测试在手机序列号上dafa已模拟SIM卡被更换的情况
20             String realsim=tm.getSimSerialNumber();
```

```
21          if(bindsim.equals(realsim)){
22              Log.i("","sim卡未发生变化,还是您的手机");
23          } else {
24              Log.i("","SIM卡变化了");
25              //由于系统版本的原因,这里的发短信可能与其他手机版本不兼容
26              String safenumber=sp.getString("safephone","");
27              if(!TextUtils.isEmpty(safenumber)){
28                  SmsManager smsManager=SmsManager.getDefault();
29                  smsManager.sendTextMessage(safenumber,null,
30                  "你的亲友手机的SIM卡已经被更换! ",null,null);
31              }
32          }
33      }
34  }
35 }
```

代码说明:

- 第9～12行代码用于获取当前手机防盗保护的状态。
- 第13～33行代码用于判断手机SIM卡是否更换,当防盗保护开启时,获取绑定的SIM卡串号,然后获取当前手机SIM卡串号进行比对,如果一致则代表SIM卡未发生变化,如果不一致则代表SIM卡发生变化,此时需要向安全号码发送短信,提示手机SIM卡已更换。

多学一招:Application 类

Application 和 Activity、Service 一样是 Android 框架的一个系统组件,当 Android 程序启动时系统会创建一个 Application 对象(只创建一个,所以 Application 可以说是单例模式的一个类),用来存储系统的一些信息,如全局变量,全局变量相对静态类更有保障,直到应用的所有 Activity 全部被销毁之后才会被释放。

通常情况下 Application 是不需要手动指定的,系统会自动创建。如果需要自己创建 Application,也很简单,只需创建一个类继承 Application,并在 AndroidManifest.xml 文件中的 application 标签进行注册即可(只需要给 application 标签增加 name 属性,并添加自己的 Application 名字)。

当 Application 启动时,系统会创建一个 PID,即进程 ID,所有的 Activity 都会在此进程上运行。也就是说,当 Application 创建时会初始化全局变量,同一个应用的所有 Activity 都可以取得这些全局变量的值,换句话说,在某一个 Activity 中改变了这些全局变量的值,那么在同一个应用的其他 Activity 中值也会改变。

Application 对象的生命周期是整个程序中最长的,它的生命周期就等于整个程序的生命周期。由于它是全局的单例的,在不同的 Activity、Service 中获得的对象都是同一个对象,因此可以通过 Application 来进行一些数据传递、数据共享和数据缓存等操作。

在 Android 系统中，有些手机 SIM 卡更换后，需要重新启动手机识别新的 SIM 卡，因此，为了最大限度地知道 SIM 卡变化，还需要创建一个开机启动的广播接收者，监听手机开机事件，并调用 App 中的 correctSIM()方法判断 SIM 卡是否变更，开机启动的广播接收者如【文件 2-31】所示。

【文件 2-31】BootCompleteReciever.java

```java
1  /** 监听开机启动的广播接收者，主要用于检查SIM卡是否被更换，如果被更换则发送短信给安全号码 */
2  public class BootCompleteReciever extends BroadcastReceiver{
3      private static final String TAG = BootCompleteReciever.class.getSimpleName();
4      @Override
5      public void onReceive(Context context, Intent intent){
6          ((App) context.getApplicationContext()).correctSIM();  //初始化
7      }
8  }
```

在 AndroidManifest.xml 文件中，注册开机启动的广播接收者以及配置权限信息，具体代码如下所示：

```xml
<uses-permission android:name="android.permission.SEND_SMS"/>
<uses-permission android:name="android.permission.RECEIVE_BOOT_COMPLETED"/>
<application
    android:name="cn.itcast.mobliesafe.App"
    android:allowBackup="true"
    android:icon="@drawable/ic_launcher"
    android:label="@string/app_name"
    android:theme="@style/AppTheme">
    <receiver android:name="cn.itcast.mobliesafe.chapter02.receiver.BootCompleteReciever">
        <intent-filter>
            <action android:name="android.intent.action.BOOT_COMPLETED"/>
        </intent-filter>
    </receiver>
</application >
```

2.5.5 向导功能（三）

设置向导第三个界面用于选择或输入安全联系人，当手机 SIM 卡变更后会向安全号码发送短信通知，具体如【文件 2-32】所示。

【文件 2-32】SetUp3Activity.java

```java
1  @SuppressLint("ShowToast")
2  public class SetUp3Activity extends BaseSetUpActivity implements OnClickListener{
3      private EditText mInputPhone;
```

```java
4    @Override
5    protected void onCreate(Bundle savedInstanceState){
6        super.onCreate(savedInstanceState);
7        setContentView(R.layout.activity_setup3);
8        initView();
9    }
10   /**
11    * 初始化控件
12    */
13   private void initView(){
14       ((RadioButton)findViewById(R.id.rb_third)).setChecked(true);
15       findViewById(R.id.btn_addcontact).setOnClickListener(this);
16       mInputPhone=(EditText) findViewById(R.id.et_inputphone);
17       String safephone=sp.getString("safephone",null);
18       if(!TextUtils.isEmpty(safephone)){
19           mInputPhone.setText(safephone);
20       }
21   }
22   @Override
23   public void showNext(){
24       //判断文本输入框中是否有电话号码
25       String safePhone = mInputPhone.getText().toString().trim();
26       if(TextUtils.isEmpty(safePhone)){
27           Toast.makeText(this,"请输入安全号码",0).show();
28           return;
29       }
30       Editor edit=sp.edit();
31       edit.putString("safephone", safePhone);
32       edit.commit();
33       startActivityAndFinishSelf(SetUp4Activity.class);
34   }
35   @Override
36   public void showPre(){
37       startActivityAndFinishSelf(SetUp2Activity.class);
38   }
39   @Override
40   public void onClick(View v){
41       switch (v.getId()){
42           case R.id.btn_addcontact:
```

```
43            startActivityForResult(new Intent(this,ContactSelectActivity.class),0);
44            break;
45        }
46    }
47    @Override
48    protected void onActivityResult(int requestCode,int resultCode,
49  Intent data){
50        super.onActivityResult(requestCode,resultCode,data);
51        if(data!=null){
52            String phone=data.getStringExtra("phone");
53            mInputPhone.setText(phone);
54        }
55    }
56 }
```

代码说明：

- 第 13～21 行的 initView()方法用于初始化控件，指定第三个小圆点被选中，并判断是否指定过安全号码，如果指定过则会显示在当前文本编辑框中。
- 第 23～34 行的 showNext()方法用于判断文件编辑框中是否有输入了安全号码，如果没有则弹出提示让用户输入号码，否则将号码保存在 SharedPreferences 对象中，然后开启 SetUp4Activity 进入下一个界面，并关闭当前 Activity。
- 第 40～46 行的 onClick()方法用于响应按钮的点击事件，当点击添加联系人按钮时，会跳转到 ContactSelectActivity 中选择要添加的安全号码（查询联系人功能会在下一小节实现）。
- 第 48～55 行的 onActivityResult()方法用于接收联系人界面中选中的联系人信息，并将联系人号码展示到安全号码中。

2.5.6 获取联系人

在设置向导第三个界面中，当点击"添加联系人"按钮时，需要跳转到联系人列表界面。获取联系人功能相对来说比较独立，因此将其作为一个单独小节进行讲解。

1. 联系人列表 UI

由于联系人信息是以条目依次展示的，因此，在界面中可以使用 ListView 控件，联系人列表的图形化界面如图 2-16 所示。

图 2-16 所示联系人列表界面对应的布局文件如【文件 2-33】所示。

【文件 2-33】activity_contact_select.xml

```xml
<?xml version="1.0" encoding="utf-8"?>
<LinearLayout xmlns:android="http://schemas.android.com/apk/res/android"
    android:layout_width="match_parent"
    android:layout_height="match_parent"
```

```
    android:orientation="vertical" >
    <include layout="@layout/titlebar"/>
    <ListView
        android:id="@+id/lv_contact"
        android:layout_width="match_parent"
        android:layout_height="wrap_content"
        android:cacheColorHint="#00000000"
        android:divider="#FFFFFF"
        android:dividerHeight="1dp"/>
</LinearLayout>
```

上述布局文件中，引入了一个 titlebar.xml 布局，该布局是一个标题栏，用于展示"选择联系人"的标题，以及在左侧放置一个返回按钮，当点击返回按钮时返回到设置向导第三个界面。

下面开发联系人列表的 Item 布局，用于填充 activity_contact_select.xml 布局，Item 布局的图形化界面如图 2-17 所示。

图 2-17 中 Item 对应的布局文件如【文件 2-34】所示。

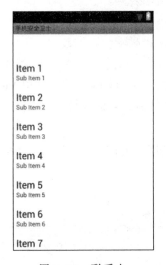 

图 2-16　联系人　　　　　图 2-17　Item 布局

【文件 2-34】item_list_contact_select.xml

```
<?xml version="1.0" encoding="utf-8"?>
<RelativeLayout xmlns:android="http://schemas.android.com/apk/res/android"
    android:layout_width="match_parent"
    android:layout_height="wrap_content"
    android:background="#10000000"
    android:orientation="vertical">
    <View
```

```
            android:id="@+id/view1"
            android:layout_width="60dp"
            android:layout_height="60dp"
            android:layout_centerVertical="true"
            android:layout_margin="15dp"
            android:background="@drawable/contact_icon"/>
        <TextView
            android:id="@+id/tv_name"
            style="@style/textview16sp"
            android:layout_alignTop="@+id/view1"
            android:layout_toRightOf="@+id/view1"
            android:textColor="@color/purple"/>
        <TextView
            android:id="@+id/tv_phone"
            style="@style/textview14sp"
            android:layout_toRightOf="@+id/view1"
            android:layout_below="@+id/tv_name"
            android:layout_marginTop="10dp"
            android:textColor="@color/purple"/>
    </RelativeLayout>
```

上述布局文件中，定义了一个 View 控件以及两个 TextView 控件，其中 View 控件用于显示联系人图标，TextView 控件分别用于显示联系人姓名、电话号码。

2. 联系人的实体类

接下来开发一个 ContactInfo 类，该类用于封装联系人信息，如姓名、手机号码等，具体代码如【文件 2-35】所示。

【文件 2-35】ContactInfo.java

```
1 public class ContactInfo{
2     public String id;
3     public String name;
4     public String phone;
5 }
```

3. 解析联系人

联系人信息都存储在 SQLite 数据库中，因此需要先获取到联系人的 id，根据 id 在 data 表中查询联系人名字以及电话号码，并封装到 ContactInfo 中，然后存入 List 集合，具体代码如【文件 2-36】所示。

【文件 2-36】ContactInfoParser.java

```
1 public class ContactInfoParser{
2     public static List<ContactInfo>getSystemContact(Context context){
```

```java
    ContentResolver resolver=context.getContentResolver();
    //1.查询raw_contacts表,把联系人的id取出来
    Uri uri=Uri.parse("content://com.android.contacts/raw_contacts");
    Uri datauri=Uri.parse("content://com.android.contacts/data");
    List<ContactInfo> infos=new ArrayList<ContactInfo>();
    Cursor cursor=resolver.query(uri, new String[] {"contact_id"},
    null,null,null);
    while(cursor.moveToNext()){
        String id=cursor.getString(0);
        if(id!=null){
            System.out.println("联系人id: "+id);
            ContactInfo info=new ContactInfo();
            info.id=id;
            //2.根据联系人的id,查询data表,把这个id的数据取出来
            //系统API查询data表的时候,不是真正的查询data表,而是查询data表的视图
            Cursor dataCursor=resolver.query(datauri,new String[]{
                "data1","mimetype"},"raw_contact_id=?",
                new String[] {id},null);
            while(dataCursor.moveToNext()){
                String data1=dataCursor.getString(0);
                String mimetype=dataCursor.getString(1);
                if("vnd.android.cursor.item/name".equals(mimetype)){
                    System.out.println("姓名="+data1);
                    info.name=data1;
                } else if("vnd.android.cursor.item/phone_v2"
                    .equals(mimetype)){
                    System.out.println("电话="+data1);
                    info.phone=data1;
                }
            }
            infos.add(info);
            dataCursor.close();
        }
    }
    cursor.close();
    return infos;
}
public static List<ContactInfo> getSimContacts( Context context){
    Uri uri = Uri.parse("content://icc/adn");
    List<ContactInfo> infos = new ArrayList<ContactInfo>();
```

```
43    Cursor mCursor = context.getContentResolver().query(uri, null, null, null, null);
44    if (mCursor != null) {
45        while (mCursor.moveToNext()) {
46            ContactInfo info = new ContactInfo();
47            // 取得联系人名字
48            int nameFieldColumnIndex = mCursor.getColumnIndex("name");
49            info.name = mCursor.getString(nameFieldColumnIndex);
50            // 取得电话号码
51            int numberFieldColumnIndex = mCursor
52                .getColumnIndex("number");
53            info.phone= mCursor.getString(numberFieldColumnIndex);
54            infos.add(info);
55        }
56    }
57    mCursor.close();
58    return infos;
59  }
60 }
```

4. 数据适配器

从数据库查询出的联系人信息，需要通过数据适配器填充到 ListView 中，接下来定义联系人列表的数据适配器 ContactAdapter，具体代码如【文件 2-37】所示。

【文件 2-37】ContactAdapter.java

```
1  public class ContactAdapter extends BaseAdapter{
2      private List<ContactInfo> contactInfos;
3      private Context context;
4      public ContactAdapter(List<ContactInfo> contactInfos,Context context){
5          super();
6          this.contactInfos=contactInfos;
7          this.context=context;
8      }
9      @Override
10     public int getCount(){
11         //TODO Auto-generated method stub
12         return contactInfos.size();
13     }
14     @Override
15     public Object getItem(int position){
16         //TODO Auto-generated method stub
17         return contactInfos.get(position);
18     }
```

```
19    @Override
20    public long getItemId(int position){
21        //TODO Auto-generated method stub
22        return position;
23    }
24    @Override
25    public View getView(int position,View convertView,ViewGroup parent){
26        ViewHolder holder=null;
27        if(convertView==null){
28            convertView= View.inflate(context,R.layout.item_list_contact_
29            select,null);
30            holder=new ViewHolder();
31            holder.mNameTV=(TextView) convertView.findViewById(R.id.tv_name);
32            holder.mPhoneTV=(TextView) convertView.findViewById(R.id.tv_phone);
33            convertView.setTag(holder);
34        }else{
35            holder=(ViewHolder) convertView.getTag();
36        }
37        holder.mNameTV.setText(contactInfos.get(position).name);
38        holder.mPhoneTV.setText(contactInfos.get(position).phone);
39        return convertView;
40    }
41    class ViewHolder{
42        TextView mNameTV;
43        TextView mPhoneTV;
44    }
45 }
```

多学一招：ListView 优化

在使用 ListView 控件的过程中，由于加载条目过多在滑动时可能造成卡顿。这是因为 ListView 在当前屏幕显示多少个条目，就会创建多少个对象，每一个条目都是一个对象。在滑动时，滑出屏幕的条目对象会被销毁，新加载到屏幕上的条目会创建新的对象，这样在 ListView 快速滑动时就会不断地创建对象→销毁对象→创建对象，并且每一个条目都需要加载一次布局，加载布局时会不断进行 findViewById()操作初始化控件，而布局 XML 文件是以树形结构进行加载，每次加载一个条目都需要从根节点进行初始化，这样对内存消耗也比较大，并且浪费时间。如果每个条目都有图片，图片加载的时间比较长，就会造成内存溢出异常。为此就需要对 ListView 进行优化，优化的目的是在滑动时不会重复创建对象，减少内存消耗和屏幕渲染处理。具体步骤如下：

（1）创建静态类

创建一个静态类，将需要加载的控件变量放在该静态类中，保证所有控件只创建一次对象，不会重复创建对象，具体代码如下：

> **多学一招：ListView 优化**
>
> ```
> static class ViewHolder{
>     TextView mNameTV;
>     TextView mPhoneTV;
> }
> ```
>
> （2）复用缓存 View 对象
>
> 在 Adapter 的 getView(int position,View convertView,ViewGroup parent)方法中，第二个参数 convertView 代表的就是之前滑出屏幕的条目对象。如果是第一次加载该方法，会创建新的 View 对象，如果滑动 ListView，滑动出屏幕的 View 对象会以缓存的形式存在，而 convertView 就是缓存的 View 对象，可以复用缓存该对象减少新对象的创建。在加载布局时先判断 convertView 是否存在，如果 convertView==null 说明没有缓存的 View 对象，则使用 View.inflate()方法加载布局，进行布局的初始化，否则复用缓存的 View 对象，具体代码如下：
>
> ```
> if(convertView==null){
>     convertView=View.inflate(context, R.layout.item_list_contact_select, null);
>     holder=new ViewHolder();
>     holder.mNameTV=(TextView) convertView.findViewById(R.id.tv_name);
>     holder.mPhoneTV=(TextView) convertView.findViewById(R.id.tv_phone);
>     convertView.setTag(holder);
> }else{
>     holder = (ViewHolder) convertView.getTag();
> }
> ```
>
> 需要注意的是，通常情况下 getView()方法中最后的返回值都是 View，但如果复用了 convertView，最后的返回值一定要改为 convertView，这样才会将布局显示到页面中。
>
> 至此 ListView 的优化已经完成，通过以上几个步骤便可保证 ListView 滑动时，无论有多少个条目或者滑动速度多快，ListView 只会创建一屏的条目对象，不会创建多余对象，对滑动的流畅度和内存占用都进行了优化。在进行项目开发中，都会采用该方法对 ListView 进行优化。

5. **联系人 Activity**

获取联系人所需的文件都已开发完成，接下来在 ContactSelectActivity 类中将数据填充到界面中即可，具体代码如【文件 2-38】所示。

【文件 2-38】ContactSelectActivity.java

```
1  public class ContactSelectActivity extends Activity implements OnClickListener {
2      private ListView mListView;
3      private ContactAdapter adapter;
4      private List<ContactInfo> systemContacts;
```

```java
5   Handler mHandler=new Handler(){
6       public void handleMessage(android.os.Message msg){
7           switch(msg.what){
8           case 10:
9               if(systemContacts!=null){
10                  adapter=new ContactAdapter(systemContacts,
11                      ContactSelectActivity.this);
12                  mListView.setAdapter(adapter);
13              }
14              break;
15          }
16      };
17  };
18  @Override
19  protected void onCreate(Bundle savedInstanceState){
20      super.onCreate(savedInstanceState);
21      requestWindowFeature(Window.FEATURE_NO_TITLE);
22      setContentView(R.layout.activity_contact_select);
23      initView();
24  }
25  private void initView(){
26      ((TextView) findViewById(R.id.tv_title)).setText("选择联系人");
27      ImageView mLeftImgv = (ImageView) findViewById(R.id.imgv_leftbtn);
28      mLeftImgv.setOnClickListener(this);
29      mLeftImgv.setImageResource(R.drawable.back);
30      //设置导航栏颜色
31      findViewById(R.id.rl_titlebar).setBackgroundColor(
32          getResources().getColor(R.color.purple));
33      mListView=(ListView) findViewById(R.id.lv_contact);
34      new Thread(){
35          public void run(){
36              systemContacts=ContactInfoParser.getSystemContact(
37                  ContactSelectActivity.this);
38              systemContacts.addAll(ContactInfoParser.getSimContacts(
39                  ContactSelectActivity.this));
40              mHandler.sendEmptyMessage(10);
41          };
42      }.start();
43      mListView.setOnItemClickListener(new OnItemClickListener(){
```

```
44        @Override
45        public void onItemClick(AdapterView<?> parent,View view,
46            int position,long id){
47            ContactInfo item=(ContactInfo) adapter.getItem(position);
48            Intent intent=new Intent();
49            intent.putExtra("phone",item.phone);
50            setResult(0,intent);
51            finish();
52        }
53    });
54 }
55 @Override
56 public void onClick(View v){
57 switch(v.getId()){
58    case R.id.imgv_leftbtn:
59        finish();
60        break;
61    }
62 }
63}
```

代码说明：

- 第 25~54 行的 initView()方法用于初始化控件，并注册 ListView 的条目点击事件，然后通过 Adapter 将联系人信息添加到界面中。
- 第 56~62 行的 onClick()方法用于响应返回按钮的点击事件，当点击该按钮时关闭当前窗口。

2.5.7 向导功能（四）

设置向导第四个界面用于展示设置完成界面，并默认开启防盗保护功能（可以手动关闭防盗保护），具体代码如【文件 2-39】所示。

【文件 2-39】SetUp4Activity.java

```
1 public class SetUp4Activity extends BaseSetUpActivity{
2    private TextView mStatusTV;
3    private ToggleButton mToggleButton;
4    @Override
5    protected void onCreate(Bundle savedInstanceState){
6        super.onCreate(savedInstanceState);
7        setContentView(R.layout.activity_setup4);
8        initView();
9    }
10   private void initView(){
```

```
11      ((RadioButton)findViewById(R.id.rb_four)).setChecked(true);
12      mStatusTV=(TextView) findViewById(R.id.tv_setup4_status);
13      mToggleButton=(ToggleButton)
14      findViewById(R.id.togglebtn_security_function);
15      mToggleButton.setOnCheckedChangeListener(new OnCheckedChange
16      Listener(){
17         @Override
18         public void onCheckedChanged(CompoundButton buttonView,
19         boolean isChecked){
20            if(isChecked){
21               mStatusTV.setText("防盗保护已经开启");
22            }else{
23               mStatusTV.setText("防盗保护没有开启");
24            }
25            Editor editor=sp.edit();
26            editor.putBoolean("protecting",isChecked);
27            editor.commit();
28         }
29      });
30      boolean protecting=sp.getBoolean("protecting",true);
31      if(protecting){
32         mStatusTV.setText("防盗保护已经开启");
33         mToggleButton.setChecked(true);
34      }else{
35         mStatusTV.setText("防盗保护没有开启");
36         mToggleButton.setChecked(false);
37      }
38   }
39   @Override
40   public void showNext(){
41      //跳转至防盗保护页面
42      Editor editor=sp.edit();
43      editor.putBoolean("isSetUp",true);
44      editor.commit();
45      startActivityAndFinishSelf(LostFindActivity.class);
46   }
47   @Override
48   public void showPre(){
49      startActivityAndFinishSelf(SetUp3Activity.class);
```

```
50    }
51 }
```

代码说明：

- 第 10~38 行的 initView()方法用于初始化控件，设置第四个小圆点被选中，并为 Toggle Button 按钮注册 onCheckedChanged 事件，当按钮为被选中状态时，显示 "防盗保护已经开启"，反之则显示 "防盗保护没有开启"，然后将按钮的状态存储到 SharedPreferences 对象中。其中第 27~34 行代码用于判断防盗保护是否开启，如果已经开启则将按钮设置为被选中状态，否则将按钮设置为未被选中状态。
- 第 40~46 行的 showNext()方法用于将设置过向导的状态存储到 SharedPreferences 中，然后关闭当前界面并跳转到 LostFindActivity 类中。
- 第 48~50 行的 showPre()方法用于将显示上一个界面，并关闭当前界面。

2.5.8 防盗指令

防盗指令界面可以控制防盗保护功能的开启和关闭、重新进入设置向导界面，以及展示当前可通过哪些指令来远程操控手机，具体代码如【文件 2-40】所示。

【文件 2-40】LostFindActivity.java

```
1  public class LostFindActivity extends Activity implements OnClickListener{
2      private TextView mSafePhoneTV;
3      private RelativeLayout mInterSetupRL;
4      private SharedPreferences msharedPreferences;
5      private ToggleButton mToggleButton;
6      private TextView mProtectStatusTV;
7      protected void onCreate(android.os.Bundle savedInstanceState){
8          super.onCreate(savedInstanceState);
9          requestWindowFeature(Window.FEATURE_NO_TITLE);
10         setContentView(R.layout.activity_lostfind);
11         msharedPreferences=getSharedPreferences("config",MODE_PRIVATE);
12         if(!isSetUp()){
13             //如果没有进入过设置向导，则进入
14             startSetUp1Activity();
15         }
16         initView();
17     }
18     private boolean isSetUp(){
19         return msharedPreferences.getBoolean("isSetUp",false);
20     }
21     /**初始化控件*/
22     private void initView(){
23         TextView mTitleTV=(TextView) findViewById(R.id.tv_title);
```

```java
24    mTitleTV.setText("手机防盗");
25    ImageView mLeftImgv=(ImageView) findViewById(R.id.imgv_leftbtn);
26    mLeftImgv.setOnClickListener(this);
27    mLeftImgv.setImageResource(R.drawable.back);
28    findViewById(R.id.rl_titlebar).setBackgroundColor(
29        getResources().getColor(R.color.purple));
30    mSafePhoneTV=(TextView) findViewById(R.id.tv_safephone);
31    mSafePhoneTV.setText(msharedPreferences.getString("safephone", ""));
32    mToggleButton=(ToggleButton) findViewById(R.id.togglebtn_lostfind);
33    mInterSetupRL=(RelativeLayout) findViewById(R.id.rl_inter_setup_wizard);
34    mInterSetupRL.setOnClickListener(this);
35    mProtectStatusTV=(TextView) findViewById(R.id.tv_lostfind_protectstauts);
36    //查询手机防盗是否开启，默认为开启
37    boolean protecting=msharedPreferences.getBoolean("protecting",true);
38    if(protecting){
39        mProtectStatusTV.setText("防盗保护已经开启");
40        mToggleButton.setChecked(true);
41    }else{
42        mProtectStatusTV.setText("防盗保护没有开启");
43        mToggleButton.setChecked(false);
44    }
45    mToggleButton.setOnCheckedChangeListener(new OnCheckedChange
46    Listener(){
47        @Override
48        public void onCheckedChanged(CompoundButton buttonView,boolean
49        isChecked){
50            if(isChecked){
51                mProtectStatusTV.setText("防盗保护已经开启");
52            }else{
53                mProtectStatusTV.setText("防盗保护没有开启");
54            }
55            Editor editor=msharedPreferences.edit();
56            editor.putBoolean("protecting", isChecked);
57            editor.commit();
58        }
59    });
60    }
61    @Override
62    public void onClick(View v){
```

```
63      switch (v.getId()){
64      case R.id.rl_inter_setup_wizard:
65          //重新进入设置向导
66          startSetUp1Activity();
67          break;
68      case R.id.imgv_leftbtn:
69          //返回
70          finish();
71          break;
72      }
73  }
74  private void startSetUp1Activity(){
75      Intent intent=new Intent(LostFindActivity.this,SetUp1Activity.class);
76      startActivity(intent);
77      finish();
78  }
79 }
```

代码说明：

- 第 18~20 行的 isSetUp()方法用于获取 SharedPreferences 对象中存入的 isSetUp 是否为 true，用来判断是否设置过向导。
- 第 37~44 行代码用于查询手机防盗是否开启（默认状态为开启），如果已经开启，则将文字展示为"防盗保护已经开启"，并将 ToggleButton 按钮设置为 true；如果没有开启，则将文字展示为"防盗保护没有开启"，并将 ToggleButton 按钮设置为 false。
- 第 45~60 行代码是为 ToggleButton 按钮注册状态改变的监听事件，当按钮被选中时，则当前文字显示"防盗保护已经开启"，否则显示"防盗保护没有开启"，最后将防盗保护状态是否开启存入 SharedPreferences 对象中。
- 第 62~73 行的 onClick()方法用于响应"重新进入设置向导"按钮以及标题栏返回按钮的点击事件，当点击重新进入设置向导按钮时，进入设置向导第一个界面，当点击返回按钮时，则关闭当前 Activity，返回主界面。
- 第 74~78 行的 startSetUp1Activity()方法用于跳转到 SetUp1Activity 界面重新进入设置向导，并关闭当前界面。

为了监听安全号码发送的防盗指令，需要创建一个广播接收者，根据收到的防盗指令来执行不同的操作，具体代码如【文件 2-41】所示。

【文件 2-41】SmsLostFindReciver.java

```
1 public class SmsLostFindReciver extends BroadcastReceiver{
2    private static final String TAG = SmsLostFindReciver.class.getSimpleName();
3    private SharedPreferences sharedPreferences;
4    @Override
5    public void onReceive(Context context,Intent intent){
```

```
6    sharedPreferences=context.getSharedPreferences("config",
7    Activity.MODE_PRIVATE);
8    boolean protecting=sharedPreferences.getBoolean("protecting",true);
9    if(protecting) {//防盗保护开启
10       //获取超级管理员
11       DevicePolicyManager dpm=(DevicePolicyManager) context.
12       getSystemService(Context.DEVICE_POLICY_SERVICE);
13       Object[] objs=(Object[]) intent.getExtras().get("pdus");
14       for(Object obj:objs){
15           SmsMessage smsMessage=SmsMessage.createFromPdu((byte[]) obj);
16           String sender=smsMessage.getOriginatingAddress();
17           String body=smsMessage.getMessageBody();
18           String safephone=sharedPreferences.getString("safephone",null);
19           //如果该短信是安全号码发送的
20           if(!TextUtils.isEmpty(safephone) & sender.equals(safephone)){
21               if("#*location*#".equals(body)){
22                   Log.i(TAG,"返回位置信息.");
23                   //获取位置 放在服务里面去实现。
24                   Intent service = new Intent(context,GPSLocationService. class);
25                   context.startService(service);
26                   abortBroadcast();
27               }else if("#*alarm*#".equals(body)){
28                   Log.i(TAG,"播放报警音乐.");
29                   MediaPlayer player = MediaPlayer.create(context,R.raw.ylzs);
30                   player.setVolume(1.0f,1.0f);
31                   player.start();
32                   abortBroadcast();
33               }else if("#*wipedata*#".equals(body)){
34                   Log.i(TAG,"远程清除数据.");
35                   dpm.wipeData(DevicePolicyManager.WIPE_EXTERNAL_STORAGE);
36                   abortBroadcast();
37               }else if("#*lockscreen*#".equals(body)){
38                   Log.i(TAG, "远程锁屏.");
39                   dpm.resetPassword("123",0);
40                   dpm.lockNow();    //没有管理员权限,调用时会崩溃
41                   abortBroadcast();
42               }
43           }
44       }
```

```
45        }
46    }
47 }
```

代码说明：

- 第 6~8 行代码用于获取 SharedPreferences 对象中存入的 key 为 protecting（表示防盗保护是否开启）的值，如果开启则执行 if 语句中的操作。
- 第 11~12 行代码用于获取超级管理员权限，只有超级管理员才能完成远程清除数据和远程锁屏功能，超级管理员权限需要在清单文件中配置。
- 第 13~17 行代码用于遍历数据库中的短信，获取发件人以及发送的短信内容。
- 第 18~44 行代码用于获取安全联系人的号码，并判断发送短信的号码中是否有安全联系人发送的，如果有则执行下面的 if 语句，分别根据发送的防盗指令返回位置信息、播放报警音乐、远程清除数据、远程锁屏功能。其中第 35 行清除数据代码和第 40 行锁屏代码是必须要有超级管理员权限的，并且当调用这段代码时，必须开启超级管理员权限，否则系统会崩溃。

值得一提的是，开启超级管理员权限有两种方式，一种是手动开启，另一种是通过代码开启。手动开启相对来说比较简单，通过"设置"→"安全"→"设备管理器"选中手机安全卫士项目即可。代码开启相对比较麻烦，不过在程序开发中大部分都是通过代码开启超级管理员权限。

在 Android 系统中，为了信息安全，普通用户是无法随意删除系统中数据的，因此在执行远程锁屏和删除数据时，需要获得超级管理员权限，接下来定义一个超级管理员的广播接收者，具体代码如【文件 2-42】所示。

【文件 2-42】MyDeviceAdminReciever.java

```
1 /**1.定义特殊的广播接收者,系统超级管理员的广播接收者*/
2 public class MyDeviceAdminReciever extends DeviceAdminReceiver{
3     @Override
4     public void onReceive(Context context,Intent intent){
5         //TODO Auto-generated method stub
6     }
7 }
```

在 AndroidManifest.xml 文件中，注册超级管理员的广播接收者，具体代码如【文件 2-43】所示。

【文件 2-43】AndroidManifest.xml

```
<!--2.配置设备超级管理员广播接收者,引用XML策略声明device_admin_sample-->
<receiver
    android:name="cn.itcast.mobliesafe.chapter02.receiver.MyDeviceAdminReciever"
    android:description="@string/sample_device_admin_description"
    android:label="@string/sample_device_admin"
    android:permission="android.permission.BIND_DEVICE_ADMIN">
    <meta-data
```

```
        android:name="android.app.device_admin"
        android:resource="@xml/device_admin_sample"/>
    <intent-filter>
        <action android:name="android.app.action.DEVICE_ADMIN_ENABLED"/>
    </intent-filter>
</receiver>
```

上述代码中,定义了一个描述信息 description 以及 label,这些信息是放置在 strings.xml 文件中的,还添加了一个权限 android:permission="android.permission.BIND_DEVICE_ADMIN",用于指定绑定超级管理员的权限。

\<meta-data\>标签表示超级管理员的元数据,其中的 resource 属性表示资源,用于指定安全策略,在此将安全策略放置在 device_admin_sample.xml 文件中,具体代码如【文件 2-44】所示。

【文件 2-44】res/xml/device_admin_sample.xml

```xml
<device-admin xmlns:android="http://schemas.android.com/apk/res/android" >
    <!--3.声明安全策略,锁屏、清除数据、重置密码 -->
    <uses-policies>
        <force-lock/>         <!-- 锁屏 -->
        <wipe-data/>          <!-- 清除数据 -->
        <reset-password/>     <!-- 重置密码 -->
    </uses-policies>
</device-admin>
```

下面创建防盗指令中使用的定位服务(GPSLocationService),该服务用于获取手机的经度、纬度、移动速度、精确度等,具体代码如【文件 2-45】所示。

【文件 2-45】GPSLocationService.java

```java
1  /**用于定位*/
2  public class GPSLocationService extends Service{
3      private LocationManager lm;
4      private MyListener listener;
5      @Override
6      public IBinder onBind(Intent intent){
7          return null;
8      }
9      @Override
10     public void onCreate(){
11         super.onCreate();
12         lm=(LocationManager) getSystemService(LOCATION_SERVICE);
13         listener=new MyListener();
14         //criteria 查询条件
```

```java
15      //true 只返回可用的位置提供者
16      Criteria criteria=new Criteria();
17      criteria.setAccuracy(Criteria.ACCURACY_FINE);//获取准确的位置。
18      criteria.setCostAllowed(true);//允许产生开销
19      String name=lm.getBestProvider(criteria, true);
20      System.out.println("最好的位置提供者: "+name);
21      lm.requestLocationUpdates(name,0,0,listener);
22  }
23  private class MyListener implements LocationListener{
24      @Override
25      public void onLocationChanged(Location location){
26          StringBuilder sb=new StringBuilder();
27          sb.append("accuracy:"+location.getAccuracy()+"\n");
28          sb.append("speed:"+location.getSpeed()+"\n");
29          sb.append("jingdu:"+location.getLongitude()+"\n");
30          sb.append("weidu:"+location.getLatitude()+"\n");
31          String result=sb.toString();
32          SharedPreferences sp=getSharedPreferences("config",MODE_PRIVATE);
33          String safenumber=sp.getString("safephone","");
34          SmsManager.getDefault().sendTextMessage(safenumber,null,result,
35          null,null);
36          stopSelf();
37      }
38      //当位置提供者状态发生变化时调用的方法
39      @Override
40      public void onStatusChanged(String provider,int status,Bundle extras) {
41      }
42      //当某个位置提供者可用的时候调用的方法
43      @Override
44      public void onProviderEnabled(String provider){
45      }
46      //当某个位置提供者不可用的时候调用的方法
47      @Override
48      public void onProviderDisabled(String provider){
49      }
50  }
51  @Override
52  public void onDestroy(){
53      super.onDestroy();
```

```
54        lm.removeUpdates(listener);
55        listener=null;
56    }
57 }
```

代码说明：
- 第 10~22 行的 onCreate()方法用于位置提供者，首先获取系统的位置管理器，然后通过 Criteria 对象返回可用的位置提供者，通过 lm.getBestProvider(criteria,true)获取最好的位置提供者，最后通过 lm.requestLocationUpdates(name,0,0,listener)使用位置提供者。
- 第 23~50 行定义了一个位置监听器 MyListener，并实现了位置变化的四个方法，其中 onLocationChanged()方法是当手机位置发生变化时调用，因此可以在该方法中获取手机的精确度、移动速度、纬度、经度，然后拼接成一个字符串发送给安全号码。onStatusChanged()方法表示当前位置提供者位置发生变化时调用。onProviderEnabled()方法表示当前位置提供者可用的时候调用，onProviderDisabled()方法表示当前位置提供者不可用时调用。
- 第 52~56 行 onDestroy()方法用于注销当前的位置监听器。

接下来在 AndroidManifest.xml 文件中注册 GPSLocationService 服务，具体代码如下所示：

```
<service
    android:name="cn.itcast.mobliesafe.chapter02.service.GPSLocation Service"
    android:persistent="true" >
</service>
```

2.5.9 修改 HomeActivity 文件

手机防盗模块代码已编写完成，接下来需要在 HomeActivity 中进行调用，在调用该模块代码时，首先需要调用设置密码对话框代码，在这段代码中判断两次输入的密码是否一致，是否设置过密码等，如果设置过密码则调用输入密码对话框代码，判断密码是否正确，具体代码如【文件 2-46】所示。

【文件 2-46】HomeActivity.java

```
1  public class HomeActivity extends Activity {
2      /**声明 GridView 该控件类似 ListView*/
3      private GridView gv_home;
4      /**存储手机防盗密码的 sp */
5      private SharedPreferences msharedPreferences;
6      /**设备管理员*/
7      private DevicePolicyManager policyManager;
8      /**申请权限*/
9      private ComponentName componentName;
10     private long mExitTime;
11     @Override
12     protected void onCreate(Bundle savedInstanceState){
```

```java
13      super.onCreate(savedInstanceState);
14      //初始化布局
15      requestWindowFeature(Window.FEATURE_NO_TITLE);
16      setContentView(R.layout.activity_home);
17      msharedPreferences = getSharedPreferences("config", MODE_PRIVATE);
18      //初始化 GridView
19      gv_home = (GridView) findViewById(R.id.gv_home);
20      gv_home.setAdapter(new HomeAdapter(HomeActivity.this));
21      //设置条目的点击事件
22      gv_home.setOnItemClickListener(new OnItemClickListener(){
23          /**parent 代表 gridView, view 代表每个条目的 view 对象, postion 代表
24             每个条目的位置*/
25          @Override
26          public void onItemClick(AdapterView<?> parent,View view,
27          int position,long id){
28              switch(position){
29              case 0: //点击手机防盗
30                  if(isSetUpPassword()){
31                      //弹出输入密码对话框
32                      showInterPswdDialog();
33                  } else {
34                      //弹出设置密码对话框
35                      showSetUpPswdDialog();
36                  }
37                  break;
38              case 1: //点击通讯卫士
39                  startActivity(SecurityPhoneActivity.class);
40                  break;
41              case 2: //软件管家
42                  startActivity(AppManagerActivity.class);
43                  break;
44              case 3://手机杀毒
45                  startActivity(VirusScanActivity.class);
46                  break;
47              case 4://缓存清理
48                  startActivity(CacheClearListActivity.class);
49                  break;
50              case 5://进程管理
51                  startActivity(ProcessManagerActivity.class);
```

```
52                break;
53            case 6: //流量统计
54                startActivity(TrafficMonitoringActivity.class);
55                break;
56            case 7: //高级工具
57                startActivity(AdvancedToolsActivity.class);
58                break;
59            case 8: //设置中心
60                startActivity(SettingsActivity.class);
61                break;
62            }
63        }
64    });
65    //1.获取设备管理员
66    policyManager=(DevicePolicyManager) getSystemService(DEVICE_POLICY_SERVICE);
67    //2.申请权限,MyDeviceAdminReciever 继承自 DeviceAdminReceiver
68    componentName=new ComponentName(this, MyDeviceAdminReciever.class);
69    //3.判断,如果没有权限则申请权限
70    boolean active=policyManager.isAdminActive(componentName);
71    if(!active){
72        //没有管理员的权限,则获取管理员的权限
73        Intent intent=new Intent(DevicePolicyManager.ACTION_ADD_DEVICE_ADMIN);
74        intent.putExtra(DevicePolicyManager.EXTRA_DEVICE_ADMIN,component Name);
75        intent.putExtra(DevicePolicyManager.EXTRA_ADD_EXPLANATION,
76            "获取超级管理员权限,用于远程锁屏和清除数据");
77        startActivity(intent);
78    }
79 }
80 /**
81  * 弹出设置密码对话框
82  */
83 private void showSetUpPswdDialog(){
84    final SetUpPasswordDialog setUpPasswordDialog=new SetUpPassword
85    Dialog(HomeActivity.this);
86    setUpPasswordDialog.setCallBack(new cn.itcast.mobliesafe.
87    chapter02.dialog.SetUpPasswordDialog. MyCallBack(){
88        @Override
89        public void ok(){
```

```
90          String firstPwsd=setUpPasswordDialog.mFirstPWDET.
91              getText().toString().trim();
92          String affirmPwsd=setUpPasswordDialog.mAffirmET.
93              getText().toString().trim();
94          if(!TextUtils.isEmpty(firstPwsd) &&
95              !TextUtils.isEmpty(affirmPwsd)){
96                  if(firstPwsd.equals(affirmPwsd)){
97                      //两次密码一致,存储密码
98                      savePswd(affirmPwsd);
99                      setUpPasswordDialog.dismiss();
100                     //显示输入密码对话框
101                     showInterPswdDialog();
102                 }else{
103                     Toast.makeText(HomeActivity.this,"两次密码不一致! ",
104                         0).show();
105                 }
106             }else{
107                 Toast.makeText(HomeActivity.this,"密码不能为空! ",0).show();
108             }
109         }
110             @Override
111             public void cancle(){
112                 setUpPasswordDialog.dismiss();
113             }
114         });
115     setUpPasswordDialog.setCancelable(true);
116     setUpPasswordDialog.show();
117 }
118 /**
119  * 弹出输入密码对话框
120  */
121 private void showInterPswdDialog(){
122     final String password=getPassword();
123     final InterPasswordDialog mInPswdDialog=new InterPassword Dialog(
124     HomeActivity.this);
125     mInPswdDialog.setCallBack(new MyCallBack(){
126         @Override
127         public void confirm(){
```

```
128             if(TextUtils.isEmpty(mInPswdDialog.getPassword())) {
129                 Toast.makeText(HomeActivity.this,"密码不能为空! ",0).show();
130             } else if(password.equals(MD5Utils.encode(mInPswdDialog.
131                 getPassword()))){
132                 // 进入防盗主界面
133                 mInPswdDialog.dismiss();
134                 startActivity(LostFindActivity.class);
135             } else{
136                 // 对话框消失,弹出土司
137                 mInPswdDialog.dismiss();
138                 Toast.makeText(HomeActivity.this, "密码有误,请重新输入! ", 0).show();
139             }
140         }
141         @Override
142         public void cancle(){
143             mInPswdDialog.dismiss();
144         }
145     });
146     mInPswdDialog.setCancelable(true);
147     //让对话框显示
148     mInPswdDialog.show();
149 }
150 /**
151  * 保存密码
152  * @param affirmPwsd
153  */
154 private void savePswd(String affirmPwsd){
155     Editor edit=msharedPreferences.edit();
156     //为了防止用户隐私被泄露,因此需要加密密码
157     edit.putString("PhoneAntiTheftPWD", MD5Utils.encode(affirmPwsd));
158     edit.commit();
159 }
160 /**
161  * 获取密码
162  * @return sp存储的密码
163  */
164 private String getPassword(){
165     String password=msharedPreferences.getString("PhoneAntiThe
166                     ftPWD",null);
```

```java
167     if(TextUtils.isEmpty(password)){
168         return "";
169     }
170     return password;
171 }
172 /**判断用户是否设置过手机防盗密码 */
173 private boolean isSetUpPassword(){
174     String password=msharedPreferences.getString("PhoneAntiTheftPWD", null);
175     if(TextUtils.isEmpty(password)){
176         return false;
177     }
178     return true;
179 }
180 /**
181  * 开启新的activity不关闭自己
182  * @param cls 新的activity的字节码
183  */
184 public void startActivity(Class<?> cls){
185     Intent intent=new Intent(HomeActivity.this,cls);
186     startActivity(intent);
187 }
188 /**
189  * 按两次返回键退出程序
190  */
191 @Override
192 public boolean onKeyDown(int keyCode, KeyEvent event){
193     if(keyCode==KeyEvent.KEYCODE_BACK){
194         if((System.currentTimeMillis()-mExitTime)>2000){
195             Toast.makeText(this, "再按一次退出程序", Toast.LENGTH_SHORT).show();
196             mExitTime=System.currentTimeMillis();
197         } else{
198             System.exit(0);
199         }
200         return true;
201     }
202     return super.onKeyDown(keyCode,event);
203 }
204 }
```

代码说明：

- 第 22～64 行代码用于响应 GridView 条目的点击事件，当点击第一个功能图标时，也就是 switch 语句中的 case 等于 0 时，会进入手机防盗功能模块。点击其他的图标会进入相应模块，在此将其跳转代码全部实现。
- 第 66～78 行代码用于获取设备的超级管理员权限，首先得到 PolicyManager 对象，然后通过该对象申请超级管理员权限。在程序启动时这段代码就会执行，方便后期通过管理员权限锁屏以及清除数据。
- 第 83～117 行的 showSetUpPswdDialog()方法用于弹出设置密码对话框，首先在 setUpPasswordDialog.setCallBack()回调方法中创建 SetUpPasswordDialog.MyCallBack(){}对象，实现该回调接口中的 ok()方法及 cancle()方法，并通过 if 语句判断两次输入的密码是否一致。
- 第 121～149 行的 showInterPswdDialog()方法用于弹出输入密码对话框，同样在 mInPswdDialog.setCallBack()方法中创建 InterPasswordDialog.MyCallBack(){}对象，实现 confirm()方法以及 cancle()方法，当点击"确定"按钮时判断输入的密码是否正确，当点击"取消"按钮时关闭对话框。
- 第 154～159 行的 savePswd()方法用于将加密过的密码保存到 SharedPreferences 对象中。
- 第 164～171 行的 getPassword()方法用于从 SharedPreferences 对象中获取保存的密码。
- 第 173～179 行的 isSetUpPassword()方法用于判断用户是否设置过手机防盗密码。

本 章 小 结

本章主要针对手机防盗模块进行讲解，首先针对该模块功能进行介绍，然后讲解防盗密码功能，以及设置向导 UI 界面，最后针对设置向导功能的实现进行讲解。学完该模块后会掌握手势识别、超级管理员权限的获取以及 GPS 定位的开发。该模块功能复杂，逻辑性强，编程者需要慢慢体会，并要动手完成所有代码的编写，为后面模块的开发奠定基础。

【面试精选】
1. 请问 Android 中有几种数据存储方式，每种方式有哪些特点？
2. 请问为什么要对 ListView 控件进行优化，如何优化？
扫描右方二维码，查看面试题答案！

通讯卫士模块

学习目标

- 了解通讯卫士模块功能；
- 掌握 SQLite 数据库的使用；
- 掌握如何使用广播接收者拦截电话和短信。

在日常生活中，使用手机时经常会被某些电话或短信骚扰，例如推销保险、中奖信息等，为此，我们开发了通讯卫士模块。该模块可以将骚扰电话或垃圾短信添加到黑名单中，并对其进行拦截。本章将针对通讯卫士模块进行详细讲解。

3.1 模 块 概 述

3.1.1 功能介绍

通讯卫士模块的主要功能是进行黑名单拦截，对添加到黑名单中的号码进行电话拦截或短信拦截。在添加黑名单时有两种方式，一种是直接在编辑框中输入电话号码以及联系人姓名，另一种是从联系人列表中选择。接下来分别针对这两种方式进行讲解。

1. 手动添加

当没有添加黑名单时，会展示通讯卫士主界面，此时点击"添加黑名单"按钮，会进入添加黑名单界面，该界面中可以直接输入联系人号码和姓名，选中电话拦截、短信拦截的复选框，然后点击"添加"按钮，此时会将该号码添加到黑名单数据库中，并将黑名单信息展示到通信卫士主界面中，如图 3-1 所示。

图 3-1　手动输入黑名单信息

2. 从列表选择

在添加黑名单界面中，点击"从联系人中添加"按钮时，会跳转到联系人列表界面，点击其中的任意一个联系人，此时该联系人的电话号码和姓名会显示在添加黑名单界面的编辑框中，点击下方的"添加"按钮，可将该号码添加到黑名单中，并在主界面中展示，如图 3-2 所示。

图 3-2　从联系人列表添加黑名单

3.1.2　开发流程图

通讯卫士模块的开发流程有些复杂，为了让大家更好地理解该模块的逻辑，接下来绘制一个流程图，具体如图 3-3 所示。

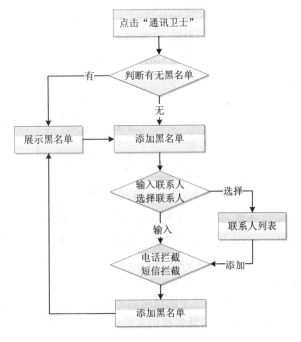

图 3-3　通讯卫士模块流程图

从图 3-3 可以看出，当进入通讯卫士界面时，首先会判断是否有黑名单，如果有则在主界面中显示黑名单列表，否则显示"您还没有添加黑名单"。添加黑名单有两种方式：一种是直接添加；另一种是从联系人列表中选择，然后指定电话拦截或者短信拦截后添加黑名单，此时黑名单信息就会展示在主界面中。

3.1.3 代码结构

通讯卫士模块代码量较大，主要逻辑包括两套主界面布局的动态切换、选择联系人、黑名单数据库的增删查、ListView 的分页查询等，通过一个图例来展示通讯卫士模块的代码结构，如图 3-4 所示。

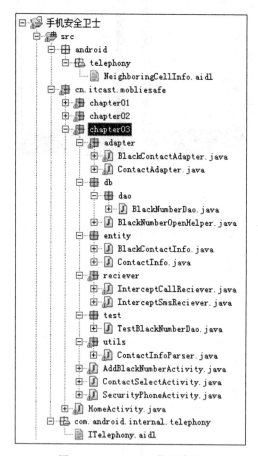

图 3-4　chapter03 代码结构

下面按照结构顺序依次介绍 chapter03 包中的文件，具体如下：
- BlackContactAdapter.java：用于填充黑名单的数据适配器；
- ContactAdapter.java：用于填充手机联系人信息的数据适配器；
- BlackNumberDao.java：用于对黑名单中的数据进行增、删、查等操作；
- BlackNumberOpenHelper.java：用于创建黑名单数据库，继承自 SQLiteOpenHelper；
- BlackContactInfo.java：用于存储黑名单信息的实体类，包括 phoneNumber、contactName 字段以及拦截模式；

- ContactInfo.java：联系人信息的实体类，包括 id、name、phone 字段；
- InterceptCallReceiver.java：拦截电话的广播接收者；
- InterceptSmsReceiver.java：拦截短信的广播接收者；
- TestBlackNumberDao.java：操作黑名单数据的测试类；
- ContactInfoParser.java：联系人信息的解析器，用于从系统通讯录中读取联系人信息；
- AddBlackNumberActivity.java：添加黑名单界面；
- ContactSelectActivity.java：显示系统联系人界面；
- SecurityPhoneActivity.java：显示黑名单信息界面。

需要注意的是，通讯卫士模块中的电话拦截功能涉及进程间通信，因此在第 3 章中引入了两个 AIDL 文件：ITelephony.aidl 和 NeighboringCellInfo.aidl，这两个文件分别位于 android.telephony 包和 com.android.internal.telephony 包中。在开发电话拦截功能时，需要创建这两个包将 AIDL 文件复制到工程中。

3.2 黑名单数据库

黑名单中的信息需要长期保存，为此可以使用 Android 自带的数据库 SQLite 存储黑名单信息。本节将针对数据库的创建以及基本操作进行详细讲解。

3.2.1 创建数据库

要想实现黑名单拦截功能，首先需要根据需求设计一个黑名单数据库（blackNumber.db），该数据库主要用于存储黑名单中的联系人信息。创建数据库代码如【文件 3-1】所示。

【文件 3-1】BlackNumberOpenHelper.java

```
1  public class BlackNumberOpenHelper extends SQLiteOpenHelper{
2      public BlackNumberOpenHelper(Context context){
3          super(context,"blackNumber.db",null,1);
4      }
5      @Override
6      public void onCreate(SQLiteDatabase db){
7        db.execSQL("create table blacknumber (id integer primary key autoincrement,
8                   number varchar(20), name varchar(255),mode integer)");
9      }
10     @Override
11     public void onUpgrade(SQLiteDatabase db,int oldVersion,int newVersion) {
12     }
13 }
```

在上述代码中，创建了一个黑名单数据库 blackNumber.db，并在数据库中创建 blacknumber 表，该表包含四个字段分别为 id、number、name、mode，其中 id 为自增主键，number 为电话号码，name 为联系人姓名，mode 为拦截模式。

3.2.2 联系人的实体类

在实现黑名单拦截时，需要保存黑名单中联系人信息，因此需要定义一个黑名单联系人的实体类，该类中包含三个属性：电话号码、联系人姓名、拦截模式，具体代码如【文件 3-2】所示。

【文件 3-2】BlackContackInfo.java

```java
1  public class BlackContactInfo{
2    /**黑名单号码*/
3    public String phoneNumber;
4    /**黑名单联系人名称*/
5    public String contactName;
6    /**黑名单拦截模式</br>   1为电话拦截   2为短信拦截   3为电话、短信都拦截*/
7    public int mode;
8    public String getModeString(int mode){
9      switch(mode){
10       case 1:
11         return "电话拦截";
12       case 2:
13         return "短信拦截";
14       case 3:
15         return "电话、短信拦截";
16     }
17     return "";
18   }
19 }
```

在上述代码的 getModeString(int mode) 方法中，定义了三种拦截模式的返回值。由于黑名单拦截模式有三种，分别为电话拦截、短信拦截以及电话短信全部拦截，因此在 getModeString(int mode) 方法中以数字 1、2、3 分别进行区分，根据传递的数字不同返回不同的拦截模式。

3.2.3 数据库操作类

由于经常添加和删除黑名单中的联系人，因此需要创建一个操作黑名单数据库的工具类，对黑名单中的数据进行增、删、查等操作，具体代码如【文件 3-3】所示。

【文件 3-3】BlackNumberDao.java

```java
1  public class BlackNumberDao{
2    private BlackNumberOpenHelper blackNumberOpenHelper;
3    public BlackNumberDao(Context context){
4      super();
5      blackNumberOpenHelper=new BlackNumberOpenHelper(context);
```

```java
6      }
7      /**
8       * 添加数据
9       * @param blackContactInfo
10      * @return
11      */
12     public boolean add(BlackContactInfo blackContactInfo){
13         SQLiteDatabase db=blackNumberOpenHelper.getWritableDatabase();
14         ContentValues values=new ContentValues();
15         if(blackContactInfo.phoneNumber.startsWith("+86")){
16             blackContactInfo.phoneNumber=blackContactInfo.phoneNumber.
17             substring(3, blackContactInfo.phoneNumber.length());
18         }
19         values.put("number",blackContactInfo.phoneNumber);
20         values.put("name",blackContactInfo.contactName);
21         values.put("mode",blackContactInfo.mode);
22         long rowid=db.insert("blacknumber", null, values);
23         if(rowid==-1){  //插入数据不成功
24             return false;
25         }else{
26             return true;
27         }
28     }
29     /**
30      * 删除数据
31      * @param blackContactInfo
32      * @return
33      */
34     public boolean delete(BlackContactInfo blackContactInfo){
35         SQLiteDatabase db=blackNumberOpenHelper.getWritableDatabase();
36         int rownumber=db.delete("blacknumber","number=?",
37         new String[] { blackContactInfo.phoneNumber });
38         if(rownumber==0){
39             return false; //删除数据不成功
40         }else{
41             return true;
42         }
43     }
44     /**
```

```java
45      * 分页查询数据库的记录
46      * @param pagenumber  第几页,页码从第 0 页开始
47      * @param pagesize    每一个页面的大小
48      */
49     public List<BlackContactInfo> getPageBlackNumber(int pagenumber,int pagesize){
50         //得到可读的数据库
51         SQLiteDatabase db=blackNumberOpenHelper.getReadableDatabase();
52         Cursor cursor = db.rawQuery("select number,mode,name from blacknumber
53                 limit ? offset ?",new String[]{String.valueOf
54                 (pagesize), String.valueOf(pagesize * pagenumber)});
55         List<BlackContactInfo> mBlackContactInfos=new ArrayList<Black
56  ContactInfo>();
57         while (cursor.moveToNext()){
58            BlackContactInfo info=new BlackContactInfo();
59            info.phoneNumber=cursor.getString(0);
60            info.mode=cursor.getInt(1);
61            info.contactName=cursor.getString(2);
62            mBlackContactInfos.add(info);
63         }
64         cursor.close();
65         db.close();
66         SystemClock.sleep(30);
67         return mBlackContactInfos;
68     }
69     /**
70      * 判断号码是否在黑名单中存在
71      * @param number
72      * @return
73      */
74     public boolean isNumberExist(String number){
75         //得到可读的数据库
76         SQLiteDatabase db=blackNumberOpenHelper.getReadableDatabase();
77         Cursor cursor=db.query("blacknumber",null,"number=?",
78             new String[] { number }, null, null, null);
79         if(cursor.moveToNext()){
80            cursor.close();
81            db.close();
82            return true;
83         }
```

```java
84          cursor.close();
85          db.close();
86          return false;
87      }
88      /**
89       * 根据号码查询黑名单信息
90       * @param number
91       * @return
92       */
93      public int getBlackContactMode(String number){
94          //得到可读的数据库
95          SQLiteDatabase db=blackNumberOpenHelper.getReadableDatabase();
96          Cursor cursor=db.query("blacknumber", new String[]{"mode"}, "number=?",
97          new String[] { number }, null, null, null);
98          int mode=0;
99          if(cursor.moveToNext()){
100                 mode=cursor.getInt(cursor.getColumnIndex("mode"));
101         }
102         cursor.close();
103         db.close();
104         return mode;
105     }
106     /**
107      * 获取数据库的总条目个数
108      * @param pagenumber 第几页,页码从第0页开始
109      * @param pagesize 每一个页面的大小
110      */
111     public int getTotalNumber(){
112         //得到可读的数据库
113         SQLiteDatabase db=blackNumberOpenHelper.getReadableDatabase();
114         Cursor cursor=db.rawQuery("select count(*) from blacknumber", null);
115         cursor.moveToNext();
116         int count=cursor.getInt(0);
117         cursor.close();
118         db.close();
119         return count;
120     }
121 }
```

代码说明:

- 第 12~28 行的 add()方法用于向数据库中添加数据,首先获取到数据库对象 SQLiteDatabase,然后创建 ContentValues 对象,并通过 if 语句判断黑名单号码是否是 "+86" 开头,如果是则对电话号码进行截取将 "+86" 去掉,最后将电话号码、姓名、拦截模式存入到数据库中,并返回数据是否插入成功的状态。
- 第 34~43 行的 delete()方法用于删除数据,同样先获取数据库对象 SQLiteDatabase 然后调用 delete()方法,根据电话号码删除数据,并返回删除结果。
- 第 49~68 行的 getPageBlackNumber()方法用于分页查询数据库中的黑名单数据。
- 第 74~87 行的 IsNumberExist()方法根据接收的号码判断当前号码是否在黑名单数据库中。
- 第 93~105 行的 getBlackContactMode()方法是根据传递进来的号码获取黑名单拦截模式,判断是电话拦截、短信拦截或者电话和短信都拦截。
- 第 111~120 行的 getTotalNumber()方法用于获取数据库中存储数据的总条目。

3.2.4 测试数据

在程序开发中,开发者需要对每一个新模块或者方法进行测试,以保证代码可运行没有 BUG。由于数据库工具类中操作黑名单数据的方法比较多,而且这些数据需要填充到主界面中,为了避免后期出现错误导致调试困难,最好在使用这些方法之前进行测试。

Android 系统自带了测试框架 JUnit,接下来使用该框架对数据库工具类中的方法进行测试,首先在清单文件中配置相应信息,具体代码如【文件 3-4】所示。

【文件 3-4】AndroidManifest.xml

```
<?xml version="1.0" encoding="utf-8"?>
<manifest xmlns:android="http://schemas.android.com/apk/res/android"
    package="cn.itcast.mobliesafe"
    android:versionCode="1"
    android:versionName="1.0">
    <uses-sdk
        android:minSdkVersion="15"
        android:targetSdkVersion="17"/>
    <instrumentation
        android:name="android.test.InstrumentationTestRunner"
        android:targetPackage="cn.itcast.mobliesafe"/>
    <application
        android:allowBackup="true"
        android:icon="@drawable/ic_launcher"
        android:label="@string/app_name"
        android:theme="@style/AppTheme">
        <uses-library android:name="android.test.runner"/>
```

```
        ...
    </application>
</manifest>
```

接下来创建一个测试类TestBlackNumberDao继承自AndroidTestCase,并对数据库中增加、删除、分页查询等方法进行测试。需要注意的是,在JUnit测试框架中,测试方法的异常必须抛出,不能try-catch,否则测试框架捕获不到异常。测试类的代码如【文件3-5】所示。

【文件3-5】TestBlackNumberDao.java

```
1  public class TestBlackNumberDao extends AndroidTestCase{
2      private Context context;
3      @Override
4      protected void setUp() throws Exception{
5          context=getContext();
6          super.setUp();
7      }
8      /**
9       * 测试添加
10      * @throws Exception
11      */
12     public void testAdd() throws Exception{
13         BlackNumberDao dao=new BlackNumberDao(context);
14         Random random=new Random(8979);
15         for(long i=0;i<30;i++){
16             BlackContactInfo info=new BlackContactInfo();
17             info.phoneNumber=135000000001+i+"";
19             info.contactName="zhangsan"+i;
19             info.mode=random.nextInt(3)+1;
20             dao.add(info);
21         }
22     }
23     /**
24      * 测试删除
25      * @throws Exception
26      */
27     public void testDelete() throws Exception{
28         BlackNumberDao dao=new BlackNumberDao(context);
29         BlackContactInfo info=new BlackContactInfo();
30         for(long i=1;i<5;i++){
31             info.phoneNumber=135000000001+i+"";
32             dao.delete(info);
```

```
33          }
34      }
35      /**
36       * 测试分页查询
37       * @throws Exception
37       */
39      public void testGetPageBlackNumber() throws Exception{
40          BlackNumberDao dao=new BlackNumberDao(context);
41          List<BlackContactInfo> list=dao.getPageBlackNumber(2,5);
42          for(int i=0;i<list.size();i++){
43              Log.i("TestBlackNumberDao",list.get(i).phoneNumber);
44          }
45      }
46      /**
47       * 测试根据号码查询黑名单信息
48       * @throws Exception
49       */
50      public void testGetBlackContactMode() throws Exception{
51          BlackNumberDao dao=new BlackNumberDao(context);
52          int mode=dao.getBlackContactMode(1350000000081+"");
53          Log.i("TestBlackNumberDao", mode+"");
54      }
55      /**
56       * 测试数据总条目
57       * @throws Exception
58       */
59      public void testGetTotalNumber() throws Exception{
60          BlackNumberDao dao=new BlackNumberDao(context);
61          int total=dao.getTotalNumber();
62          Log.i("TestBlackNumberDao","数据总条目:   "+total);
63      }
64      /**
65       * 测试号码是否在数据库中
66       * @throws Exception
67       */
68      public void testIsNumberExist() throws Exception{
69          BlackNumberDao dao=new BlackNumberDao(context);
70          boolean isExist=dao.IsNumberExist(1350000000081+"");
71          if(isExist){
```

```
72          Log.i("TestBlackNumberDao","该号码在数据库中");
73       } else {
74          Log.i("TestBlackNumberDao",
75                 "该号码不在数据库中");
76       }
77    }
78 }
```

在测试数据时，首先要测试添加方法，向数据库中添加 30 条数据，然后再测试其他方法，如果测试通过，则在 JUnit 窗口中显示绿条，如图 3-5 所示。

值得一提的是，如果在测试这几个方法时，JUnit 窗口的测试结果显示绿条，说明数据库逻辑没问题测试通过，可以正常向数据库中添加、删除或者查看数据。

图 3-5 测试成功

3.3 主 界 面

3.3.1 主界面 UI

当数据库搭建好之后，开始编写通讯卫士主界面，该界面有两种展示形式：一种是当数据库中没有黑名单数据时，显示默认图片和文字提示；另一种是当数据库中有黑名单数据时，将黑名单信息以 ListView 的方式展示在界面中。主界面的图形化界面如图 3-6 所示。

 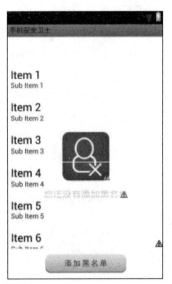

图 3-6 主界面

需要注意的是，图 3-6 中右侧黑名单列表的帧布局是隐藏的，只有在黑名单中有数据时才显示，这里为了让编程者看到更直观的效果，故将 ListView 设置为显示状态。主界面布局代码如【文件 3-6】所示。

【文件 3-6】 activity_securityphone.xml

```xml
<?xml version="1.0" encoding="utf-8"?>
<LinearLayout xmlns:android="http://schemas.android.com/apk/res/android"
    android:layout_width="match_parent"
    android:layout_height="match_parent"
    android:orientation="vertical">
    <include layout="@layout/titlebar"/>
    <RelativeLayout
        android:layout_width="match_parent"
        android:layout_height="match_parent"
        android:layout_weight="10" >
        <FrameLayout
            android:id="@+id/fl_noblacknumber"
            android:layout_width="match_parent"
            android:layout_height="match_parent">
            <RelativeLayout
                android:layout_width="match_parent"
                android:layout_height="match_parent">
                <ImageView
                    android:id="@+id/noblacknumbericon"
                    android:layout_width="100dp"
                    android:layout_height="100dp"
                    android:layout_centerInParent="true"
                    android:background="@drawable/no_blacknumbericon"/>
                <TextView
                    style="@style/textview16sp"
                    android:layout_below="@+id/noblacknumbericon"
                    android:layout_centerHorizontal="true"
                    android:layout_marginTop="16dp"
                    android:text="您还没有添加黑名单"
                    android:textColor="@color/light_gary"
                    android:textScaleX="1.2"/>
            </RelativeLayout>
        </FrameLayout>
        <FrameLayout
            android:id="@+id/fl_haveblacknumber"
            android:layout_width="match_parent"
            android:layout_height="wrap_content"
            android:visibility="gone">
```

```xml
            <ListView
                android:id="@+id/lv_blacknumbers"
                android:layout_width="match_parent"
                android:layout_height="wrap_content"
                android:dividerHeight="1dp"
                android:divider="#ffffff"/>
        </FrameLayout>
    </RelativeLayout>
    <LinearLayout
        android:layout_width="match_parent"
        android:layout_height="wrap_content"
        android:gravity="center" >
        <Button
            android:id="@+id/btn_addblacknumber"
            android:layout_width="170dp"
            android:layout_height="40dp"
            android:layout_gravity="center"
            android:layout_margin="10dp"
            android:background="@drawable/add_blacknumber_btn_selector"/>
    </LinearLayout>
</LinearLayout>
```

在上述代码中，主要包含两个 FrameLayout 布局，这两个布局分别用于显示有黑名单和无黑名单的情况，当这两个布局在切换时实际上就是控制两个 FrameLayout 的显示与隐藏。当前布局的最下方有一个 Button 按钮，该按钮的背景图片使用了图片选择器，在按钮按下与松开时显示不同颜色的图片，图片选择器的代码如【文件 3-7】所示。

【文件 3-7】res/ drawable/ add_blacknumber_btn_selector.xml

```xml
<?xml version="1.0" encoding="utf-8"?>
<selector xmlns:android="http://schemas.android.com/apk/res/android">
    <item android:state_pressed="true"
        android:drawable="@drawable/add_blackname_btn_p"/>
    <item android:state_pressed="false"
        android:drawable="@drawable/add_blackname_btn_n"/>
</selector>
```

从上述代码可以看出，当按钮处于按下状态时，使用紫色的背景图片（add_blackname_btn_p.png），当按钮弹起时，使用灰色的背景图片（add_blackname_btn_n.png）。

3.3.2 黑名单 Item 布局

由于主界面的黑名单列表是通过 ListView 控件展示的，因此需要定义一个黑名单的 Item

布局，保证每个条目的布局都是一致的，Item 布局的图形化界面如图 3-7 所示。

图 3-7 所示黑名单 Item 对应的布局文件如【文件 3-8】所示。

【文件 3-8】item_list_blackcontact.xml

图 3-7 黑名单 Item 布局

```xml
<?xml version="1.0" encoding="utf-8"?>
<RelativeLayout xmlns:android="http://schemas.
    android.com/apk/res/android"
    android:layout_width="match_parent"
    android:layout_height="wrap_content"
    android:background="#10000000"
    android:orientation="vertical" >
    <View
        android:id="@+id/view_black_icon"
        android:layout_width="60dp"
        android:layout_height="60dp"
        android:layout_centerVertical="true"
        android:layout_margin="15dp"
        android:background="@drawable/brightpurple_contact_icon" />
    <TextView
        android:id="@+id/tv_black_name"
        style="@style/textview14sp"
        android:layout_alignTop="@+id/view_black_icon"
        android:layout_toRightOf="@+id/view_black_icon"
        android:textColor="@color/purple" />
    <TextView
        android:id="@+id/tv_black_mode"
        style="@style/textview14sp"
        android:layout_toRightOf="@+id/view_black_icon"
        android:layout_below="@+id/tv_black_name"
        android:layout_marginTop="10dp"
        android:textColor="@color/purple" />
    <View
        android:id="@+id/view_black_delete"
        android:layout_width="34dp"
        android:layout_height="40dp"
        android:background="@drawable/delete_icon"
        android:layout_alignParentRight="true"
        android:layout_marginRight="20dp"
        android:layout_centerVertical="true"/>
</RelativeLayout>
```

上述布局中，使用了两个 View 对象以及两个 TextView 控件，其中 View 控件用于表示联系人图标和删除图标，TextView 控件用于显示联系人号码以及拦截方式。

3.3.3 主界面逻辑代码

主界面的功能主要包含显示黑名单信息（姓名、电话、拦截模式），点击"删除"按钮将当前黑名单从界面及数据库中删除。接下来编写主界面业务逻辑，具体代码如【文件 3-9】所示。

【文件 3-9】 SecurityPhoneActivity.java

```java
1  public class SecurityPhoneActivity extends Activity implements OnClickListener {
2    /** 有黑名单时，显示的帧布局 */
3    private FrameLayout mHaveBlackNumber;
4    /** 没有黑名单时，显示的帧布局 */
5    private FrameLayout mNoBlackNumber;
6    private BlackNumberDao dao;
7    private ListView mListView;
8    private int pagenumber=0;
9    private int pagesize=15;
10   private int totalNumber;
11   private List<BlackContactInfo> pageBlackNumber=new ArrayList<Black ContactInfo>();
12   private BlackContactAdapter adapter;
13   @Override
14   protected void onCreate(Bundle savedInstanceState){
15     super.onCreate(savedInstanceState);
16     requestWindowFeature(Window.FEATURE_NO_TITLE);
17     setContentView(R.layout.activity_securityphone);
18     initView();
19     fillData();
20   }
21   @Override
22   protected void onResume(){
23     super.onResume();
24     if(totalNumber!=dao.getTotalNumber()){
25       //数据发生变化
26       if(dao.getTotalNumber()>0){
27         mHaveBlackNumber.setVisibility(View.VISIBLE);
28         mNoBlackNumber.setVisibility(View.GONE);
29       } else {
30         mHaveBlackNumber.setVisibility(View.GONE);
31         mNoBlackNumber.setVisibility(View.VISIBLE);
```

```
32          }
33          pagenumber=0;
34          pageBlackNumber.clear();
35          pageBlackNumber.addAll(dao.getPageBlackNumber(pagenumber,pagesize));
36          if(adapter!=null){
37              adapter.notifyDataSetChanged();
38          }
39       }
40  }
41    /**
42     * 用于填充数据,重新刷新界面
43     */
44  private void fillData(){
45      dao=new BlackNumberDao(SecurityPhoneActivity.this);
46      totalNumber=dao.getTotalNumber();
47      if(totalNumber==0){
48          //数据库中没有黑名单数据
49          mHaveBlackNumber.setVisibility(View.GONE);
50          mNoBlackNumber.setVisibility(View.VISIBLE);
51      }else if(totalNumber>0){
52          //数据库中含有黑名单数据
53          mHaveBlackNumber.setVisibility(View.VISIBLE);
54          mNoBlackNumber.setVisibility(View.GONE);
55          pagenumber=0;
56          if(pageBlackNumber.size()>0){
57              pageBlackNumber.clear();
58          }
59          pageBlackNumber.addAll(dao.getPageBlackNumber(pagenumber,pagesize));
60          if(adapter==null){
61             adapter=new BlackContactAdapter(pageBlackNumber,
62             SecurityPhoneActivity.this);
63             adapter.setCallBack(new BlackConactCallBack(){
64                @Override
65                public void DataSizeChanged(){
66                    fillData();
67                }
68             });
69             mListView.setAdapter(adapter);
70          } else {
```

```java
71            adapter.notifyDataSetChanged();
72        }
73    }
74 }
75 /**
76  * 对控件进行初始化
77  */
78 private void initView(){
79     findViewById(R.id.rl_titlebar).setBackgroundColor(
80     getResources().getColor(R.color.bright_purple));
81     ImageView mLeftImgv=(ImageView) findViewById(R.id.imgv_leftbtn);
82     ((TextView) findViewById(R.id.tv_title)).setText("通讯卫士");
83     mLeftImgv.setOnClickListener(this);
84     mLeftImgv.setImageResource(R.drawable.back);
85     mHaveBlackNumber=(FrameLayout) findViewById(R.id.fl_haveblacknumber);
86     mNoBlackNumber=(FrameLayout) findViewById(R.id.fl_noblacknumber);
87     findViewById(R.id.btn_addblacknumber).setOnClickListener(this);
88     mListView=(ListView) findViewById(R.id.lv_blacknumbers);
89     mListView.setOnScrollListener(new OnScrollListener(){
90         @Override
91         public void onScrollStateChanged(AbsListView view,int scrollState){
92             switch(scrollState){
93             case OnScrollListener.SCROLL_STATE_IDLE: //没有滑动的状态
94                 //获取最后一个可见条目
95                 int lastVisiblePosition=mListView.getLastVisiblePosition();
96                 //如果当前条目是最后一个,则查询更多的数据
97                 if(lastVisiblePosition==pageBlackNumber.size()-1){
98                     pagenumber++;
99                     if(pagenumber*pagesize>=totalNumber){
100                        Toast.makeText(SecurityPhoneActivity.this,
101                        "没有更多的数据了",0).show();
102                    }else{
103                        pageBlackNumber.addAll(dao.getPageBlackNumber(
104                        pagenumber, pagesize));
105                        adapter.notifyDataSetChanged();
106                    }
107                }
108                break;
109            }
```

```
110        }
111        @Override
112        public void onScroll(AbsListView view,int firstVisibleItem,
113        int visibleItemCount, int totalItemCount){
114        }
115    });
116 }
117 @Override
118 public void onClick(View v){
119    switch(v.getId()){
120    case R.id.imgv_leftbtn:
121        finish();
122        break;
123    case R.id.btn_addblacknumber:
124        //跳转至添加黑名单页面
125        startActivity(new Intent(this,AddBlackNumberActivity.class));
126        break;
127    }
128 }
129}
```

代码说明：

- 第22~40行代码重写了onResume()方法，当Activity回到前台时调用，在该方法中判断数据库总条目是否发生变化，如果发生变化则将页码置为0，清空黑名单中数据重写添加，并更新数据。其中第26~32行代码判断黑名单数据库中有没有数据，如果没有数据，则显示主界面的第一种布局，如果有数据，则显示第二种布局并将数据以分页的方式显示在界面上。
- 第78~116行的initView()方法用于初始化控件，其中第89~115行代码是ListView的滑动监听事件，onScrollStateChanged()方法主要功能是获取数据库中的数据分页显示在界面上，每页显示多少数据由自己定义，当ListView向下滑动时再次加载同样条目的数据。
- 第118~128行的onClick()方法用于响应界面按钮的点击事件，当点击左上角的返回按钮时关闭当前Activity，当点击添加黑名单按钮时，跳转到添加黑名单界面。

3.3.4 数据适配器

主界面通过ListView显示黑名单列表时，使用了数据适配器BaseAdapter，接下来创建该数据适配器类，具体代码如【文件3-10】所示。

【文件3-10】BlackContactAdapter.java

```
1 public class BlackContactAdapter extends BaseAdapter{
2     private List<BlackContactInfo> contactInfos;
3     private Context context;
```

```
4      private BlackNumberDao dao;
5      private BlackConactCallBack callBack;
6      public void setCallBack(BlackConactCallBack callBack){
7          this.callBack=callBack;
8      }
9      public BlackContactAdapter(List<BlackContactInfo> systemContacts,
10     Context context){
11         super();
12         this.contactInfos=systemContacts;
13         this.context=context;
14         dao=new BlackNumberDao(context);
15     }
16     @Override
17     public int getCount(){
18         //TODO Auto-generated method stub
19         return contactInfos.size();
20     }
21     @Override
22     public Object getItem(int position){
23         //TODO Auto-generated method stub
24         return contactInfos.get(position);
25     }
26     @Override
27     public long getItemId(int position){
28         //TODO Auto-generated method stub
29         return position;
30     }
31     @Override
32     public View getView(final int position,View convertView,ViewGroup parent){
33         ViewHolder holder=null;
34         if(convertView==null){
35             convertView=View.inflate(context, R.layout.item_list_blackcontact,null);
36             holder=new ViewHolder();
37             holder.mNameTV=(TextView) convertView .findViewById(R.id.tv_black_name);
38             holder.mModeTV=(TextView) convertView.findViewById(R.id.tv_
39             black_mode);
40             holder.mContactImgv=convertView.findViewById(R.id.view_black_icon);
41             holder.mDeleteView=convertView.findViewById(R.id.view_black_ delete);
42             convertView.setTag(holder);
```

```
43      }else{
44          holder=(ViewHolder) convertView.getTag();
45      }
46      holder.mNameTV.setText(contactInfos.get(position).contactName+"("+
47      contactInfos.get(position).phoneNumber+")");
48      holder.mModeTV.setText(contactInfos.get(position).getModeString(
49      contactInfos.get(position).mode));
50      holder.mNameTV.setTextColor(context.getResources().getColor(
51      R.color.bright_purple));
52      holder.mModeTV.setTextColor(context.getResources().getColor(
53      R.color.bright_purple));
54      holder.mContactImgv.setBackgroundResource(
55      R.drawable.brightpurple_contact_icon);
56      holder.mDeleteView.setOnClickListener(new OnClickListener(){
57          @Override
58          public void onClick(View v){
59              boolean datele=dao.delete(contactInfos.get(position));
60              if(datele){
61                  contactInfos.remove(contactInfos.get(position));
62                  BlackContactAdapter.this.notifyDataSetChanged();
63                  //如果数据库中没有数据了，则执行回调函数
64                  if(dao.getTotalNumber()==0){
65                      callBack.DataSizeChanged();
66                  }
67              }else{
68                  Toast.makeText(context,"删除失败！",0).show();
69              }
70          }
71      });
72      return convertView;
73  }
74  static class ViewHolder{
75      TextView mNameTV;
76      TextView mModeTV;
77      View mContactImgv;
78      View mDeleteView;
79  }
80  public interface BlackConactCallBack{
81      void DataSizeChanged();
```

```
82    }
83 }
```

代码说明：

- 第 9～15 行的 BlackContactAdapter()方法是 Adapter 构造方法，该方法接收两个参数 List<Black ContactInfo> 和 Context。其中，List<BlackContackInfo>表示从主界面传递的黑名单数据集合，加载到页面上的数据都从该集合中取出。
- 第 56～71 行代码定义了每个条目中删除按钮的点击事件，当点击"删除"按钮后，当前条目在数据库中被删除，并刷新当前界面。
- 第 74～79 行的 ViewHolder 是一个静态内部类，作用是使 ListView 中的控件只加载一次，优化加载速度以及内存消耗。
- 第 80～82 行的 BlackConactCallBack 是一个回调接口，该接口的实现方法在主界面 SecurityPhoneActivity 中。在回调的实现方法中调用了主界面 fillData()方法重新刷新布局页面。

需要注意的是，"删除"按钮的点击事件中（第 65 行代码）调用了 callBack.DataSizeChanged()方法，该方法是一个回调方法，在点击删除按钮后数据库中没有数据时调用。由于只有将数据库中的数据完全删除之后才调用回调函数，因此删除最后一条数据时会显示默认布局，将 ListView 布局隐藏。

3.4 添加黑名单

3.4.1 添加黑名单界面

从功能介绍中可以看出，添加黑名单界面主要由三部分组成：第一部分包含两个 EditText，用于输入要拦截的电话号码以及姓名；第二部分包含两个 CheckBox，用于选择拦截模式；第三部分包含两个按钮，当点击"添加"按钮时，直接将输入框中的黑名单号码以及姓名添加到黑名单中，当点击"从联系人中添加"按钮时，将跳转到联系人列表界面，然后选中某个联系人进行添加。添加黑名单的图形化界面如图 3-8 所示。

图 3-8 所示添加黑名单界面对应的布局文件如【文件 3-11】所示。

图 3-8 添加黑名单

【文件 3-11】activity_add_blacknumber.xml

```xml
<?xml version="1.0" encoding="utf-8"?>
<LinearLayout xmlns:android="http://schemas.android.com/apk/res/android"
    android:layout_width="match_parent"
    android:layout_height="match_parent"
    android:orientation="vertical">
    <include layout="@layout/titlebar"/>
    <EditText
        android:id="@+id/et_balcknumber"
        android:layout_width="match_parent"
```

```xml
        android:layout_height="wrap_content"
        android:layout_marginLeft="10dp"
        android:layout_marginRight="10dp"
        android:layout_marginTop="20dp"
        android:background="@drawable/coner_white_rec"
        android:hint="请输入要添加的号码"
        android:inputType="phone"
        android:paddingLeft="5dp"
        android:textColor="@color/bright_purple"
        android:textSize="16sp"/>
    <EditText
        android:id="@+id/et_blackname"
        android:layout_width="match_parent"
        android:layout_height="wrap_content"
        android:layout_marginLeft="10dp"
        android:layout_marginRight="10dp"
        android:layout_marginTop="20dp"
        android:background="@drawable/coner_white_rec"
        android:hint="请输入名称"
        android:paddingLeft="5dp"
        android:textColor="@color/bright_purple"
        android:textSize="16sp"/>
    <LinearLayout
        android:layout_width="match_parent"
        android:layout_height="wrap_content"
        android:layout_margin="15dp"
        android:gravity="center"
        android:orientation="horizontal" >
        <CheckBox
            android:id="@+id/cb_blacknumber_tel"
            android:layout_width="wrap_content"
            android:layout_height="wrap_content"
            android:layout_weight="1"
            android:button="@drawable/checkbox_selector"
            android:checked="true"
            android:paddingLeft="40dp"
            android:text="电话拦截"
            android:textColor="@color/bright_purple"
            android:textScaleX="1.1"/>
```

```xml
<CheckBox
    android:id="@+id/cb_blacknumber_sms"
    android:layout_width="wrap_content"
    android:layout_height="wrap_content"
    android:layout_weight="1"
    android:button="@drawable/checkbox_selector"
    android:checked="true"
    android:paddingLeft="40dp"
    android:text="短信拦截"
    android:textColor="@color/bright_purple"
    android:textScaleX="1.1"
    android:textSize="16sp"/>
</LinearLayout>
<Button
  android:id="@+id/add_blacknum_btn"
  android:layout_width="170dp"
  android:layout_height="40dp"
  android:layout_gravity="center_horizontal"
  android:layout_marginTop="30dp"
  android:background="@drawable/add_btn_selector_securityphone" />
<Button
  android:id="@+id/add_fromcontact_btn"
  android:layout_width="170dp"
  android:layout_height="40dp"
  android:layout_gravity="center_horizontal"
  android:layout_marginTop="20dp"
  android:background="@drawable/add_fromcontact_btn_selector_securityphone"/>
</LinearLayout>
```

在上述代码中，四个按钮都使用了背景选择器，接下来分别介绍，具体如下：

（1）CheckBox按钮是通过背景选择器控制按钮的选中与未选中的背景图片，具体代码如【文件3-12】所示。

【文件3-12】res/drawable/checkbox_selector.xml

```xml
<?xml version="1.0" encoding="utf-8"?>
<selector xmlns:android="http://schemas.android.com/apk/res/android">
   <item android:state_checked="true"
                android:drawable="@drawable/brightpurple_checkbox_c"/>
   <item android:state_checked="false"android:drawable="@drawable/checkbox_n"/>
</selector>
```

（2）"添加"按钮的背景选择器是控制按钮按下与弹起的效果，当按钮按下时显示紫色（add_btn_p.png），当按钮弹起时显示灰色（add_btn_n.png），具体代码如【文件3-13】所示。

【文件3-13】res/drawable/add_btn_selector_securityphone.xml

```xml
<?xml version="1.0" encoding="utf-8"?>
<selector xmlns:android="http://schemas.android.com/apk/res/android" >
    <item android:state_pressed="true" android:drawable="@drawable/add_btn_p"/>
    <item android:state_pressed="false" android:drawable="@drawable/add_btn_n"/>
</selector>
```

（3）"从联系人添加"按钮的背景选择器同样是控制按钮按下与弹起的效果，当按钮按下时显示紫色（add_fromcontact_p.png），当按钮弹起时显示灰色（add_fromcontact_n.png），具体代码如【文件3-14】所示。

【文件3-14】res/drawable/add_fromcontact_btn_selector_securityphone.xml

```xml
<?xml version="1.0" encoding="utf-8"?>
<selector xmlns:android="http://schemas.android.com/apk/res/android" >
    <item android:state_pressed="true" android:drawable="@drawable/add_
        fromcontact_p"/>
    <item android:state_pressed="false" android:drawable="@drawable/add_
        fromcontact_n"/>
</selector>
```

3.4.2 添加黑名单逻辑

添加黑名单主要有两种方式，第一种是手动输入姓名和手机号码，第二种是从联系人列表中选择，当点击"从联系人中添加"按钮时，会跳转到联系人列表界面，然后选择指定的联系人，此时联系人信息会显示在添加黑名单界面，然后点击添加按钮即可。具体代码如【文件3-15】所示。

【文件3-15】AddBlackNumberActivity.java

```java
1  public class AddBlackNumberActivity extends Activity implements OnClickListener {
2      private CheckBox mSmsCB;
3      private CheckBox mTelCB;
4      private EditText mNumET;
5      private EditText mNameET;
6      private BlackNumberDao dao;
7      @Override
8      protected void onCreate(Bundle savedInstanceState){
9          super.onCreate(savedInstanceState);
10         requestWindowFeature(Window.FEATURE_NO_TITLE);
11         setContentView(R.layout.activity_add_blacknumber);
12         dao=new BlackNumberDao(this);
```

```java
13      initView();
14  }
15  private void initView(){
16      findViewById(R.id.rl_titlebar).setBackgroundColor(
17      getResources().getColor(R.color.bright_purple));
18      ((TextView) findViewById(R.id.tv_title)).setText("添加黑名单");
19      ImageView mLeftImgv=(ImageView) findViewById(R.id.imgv_leftbtn);
20      mLeftImgv.setOnClickListener(this);
21      mLeftImgv.setImageResource(R.drawable.back);
22      mSmsCB=(CheckBox) findViewById(R.id.cb_blacknumber_sms);
23      mTelCB=(CheckBox) findViewById(R.id.cb_blacknumber_tel);
24      mNumET=(EditText) findViewById(R.id.et_balcknumber);
25      mNameET=(EditText) findViewById(R.id.et_blackname);
26      findViewById(R.id.add_blacknum_btn).setOnClickListener(this);
27      findViewById(R.id.add_fromcontact_btn).setOnClickListener(this);
28  }
29  @Override
30  public void onClick(View v){
31      switch(v.getId()){
32      case R.id.imgv_leftbtn:
33          finish();
34          break;
35      case R.id.add_blacknum_btn:
36          String number=mNumET.getText().toString().trim();
37          String name=mNameET.getText().toString().trim();
38          if(TextUtils.isEmpty(number)||TextUtils.isEmpty(name)){
39              Toast.makeText(this,"电话号码和手机号不能为空!",0).show();
40              return;
41          }else{
42              //电话号码和名称都不为空
43              BlackContactInfo blackContactInfo=new BlackContactInfo();
44              blackContactInfo.phoneNumber=number;
45              blackContactInfo.contactName=name;
46              if(mSmsCB.isChecked() & mTelCB.isChecked()){
47                  //两种拦截模式都选
48                  blackContactInfo.mode=3;
49              } else if(mSmsCB.isChecked() & !mTelCB.isChecked()){
50                  // 短信拦截
51                  blackContactInfo.mode=2;
```

```
52            }else if(!mSmsCB.isChecked() & mTelCB.isChecked()){
53                //电话拦截
54                blackContactInfo.mode=1;
55            }else{
56                Toast.makeText(this,"请选择拦截模式! ",0).show();
57                return;
58            }
59            if(!dao.IsNumberExist(blackContactInfo.phoneNumber)){
60                dao.add(blackContactInfo);
61            }else{
62                Toast.makeText(this,"该号码已经被添加至黑名单",0).show();
63            }
64            finish();
65        }
66        break;
67    case R.id.add_fromcontact_btn:
68        startActivityForResult(
69        new Intent(this,ContactSelectActivity.class),0);
70        break;
71    }
72 }
73 @Override
74 protected void onActivityResult(int requestCode,int resultCode,
75 Intent data){
76    super.onActivityResult(requestCode,resultCode,data);
77    if(data!=null){
78        //获取选中的联系人信息
79        String phone=data.getStringExtra("phone");
80        String name=data.getStringExtra("name");
81        mNameET.setText(name);
82        mNumET.setText(phone);
83    }
84 }
85}
```

代码说明：
- 第 15~28 行的 initView()方法用于实现控件的初始化，并动态设置标题栏的背景颜色以及返回按钮的图片资源，并为"添加""从联系人中选择"按钮注册监听器。
- 第 30~72 行的 onClick()方法用于响应按钮的点击事件，当点击"添加"按钮时，首先会对输入框进行校验，只有电话号码和姓名不为空时，才能选择拦截模式，然后校验数据库中是否有当前号码，如果没有将其添加到黑名单中；当点击"从联系人中选择"按钮时，会跳转到联系人列表界面（ContactSelectActivity）。
- 第74~84行代码的onActivityResult()方法是接收联系人列表中选中的联系人信息，并将该信息展示到文本输入框中。

3.4.3 联系人列表

前一小节中使用的联系人列表在第2章中讲解过，因此本章可以将之前的代码复制过来直接使用（这样做的目的是保证本章代码的完整性，实际开发中直接调用这部分代码即可）。接下来列举获取联系人列表需要用到布局文件以及逻辑代码，具体如下：
- item_list_contact_select.xml：联系人列表的Item布局（不需要复制）；
- activity_contact_select.xml：联系人列表布局（不需要复制）；
- ContactInfo.java：联系人信息的实体类；
- ContactInfoParser.java：解析联系人信息的工具类；
- ContactAdapter.java：联系人列表的数据适配器；
- ContactSelectActivity.java：联系人列表的主界面逻辑（记得在AndroidManifest.xml文件中注册）。

由于第2章中联系人Item布局整体风格是蓝色的，而本章是紫色的，因此，需要在数据适配器的代码中重新指定联系人图标、联系人姓名以及电话号码的文本颜色，具体代码如【文件3-16】所示。

【文件3-16】ContactAdapter.java

```
1  public class ContactAdapter extends BaseAdapter{
2    ...
3    @Override
4    public View getView(int position,View convertView,ViewGroup parent){
5      ViewHolder holder=null;
6      if(convertView==null){
7        convertView=View.inflate(context,R.layout.item_list_contact_
8        select,null);
9        holder=new ViewHolder();
10       holder.mNameTV=(TextView) convertView.findViewById(R.id.tv_name);
11       holder.mPhoneTV=(TextView) convertView.findViewById(R.id.tv_phone);
12       holder.mContactImgv=convertView.findViewById(R.id.view1);
13       convertView.setTag(holder);
14     }else{
15       holder=(ViewHolder) convertView.getTag();
16     }
17     holder.mNameTV.setText(contactInfos.get(position).name);
18     holder.mPhoneTV.setText(contactInfos.get(position).phone);
19     holder.mNameTV.setTextColor(context.getResources().getColor(
20     R.color.bright_purple));
```

```
21        holder.mPhoneTV.setTextColor(context.getResources().
22            getColor(R.color.bright_purple));
23        holder.mContactImgv.setBackgroundResource(
24            R.drawable.brightpurple_contact_icon);
25        return convertView;
26    }
27    static class ViewHolder{
28        TextView mNameTV;
29        TextView mPhoneTV;
30        View mContactImgv;
31    }
32 }
```

3.5 黑名单拦截

在 Android 系统中，当电话或短信到来时都会产生广播，因此可以利用广播接收者将广播终止，实现黑名单拦截功能，然后将电话和短信记录删除不让其在界面中显示。本节将针对黑名单拦截功能进行详细讲解。

3.5.1 拦截短信

在进行短信拦截时，需要在广播中获取到电话号码以及短信内容，然后查询该号码是否在黑名单数据库中，如果在黑名单中，则判断是哪种拦截模式，并进行拦截。接下来创建拦截短信的广播接收者，具体代码如【文件 3-17】所示。

【文件 3-17】InterceptSmsReciever.java

```
1  public class InterceptSmsReciever extends BroadcastReceiver{
2      @Override
3      public void onReceive(Context context,Intent intent){
4          SharedPreferences mSP=context.getSharedPreferences("config",
5          Context.MODE_PRIVATE);
6          boolean BlackNumStatus=mSP.getBoolean("BlackNumStatus", true);
7          if(!BlackNumStatus){
8              //黑名单拦截关闭
9              return;
10         }
11         //如果是黑名单，则终止广播
12         BlackNumberDao dao=new BlackNumberDao(context);
13         Object[] objs=(Object[]) intent.getExtras().get("pdus");
14         for(Object obj:objs){
15             SmsMessage smsMessage=SmsMessage.createFromPdu((byte[]) obj);
```

```
16          String sender=smsMessage.getOriginatingAddress();
17          String body=smsMessage.getMessageBody();
18          if(sender.startsWith("+86")){
19             sender=sender.substring(3, sender.length());
20          }
21          int mode=dao.getBlackContactMode(sender);
22          if(mode==2||mode==3){
23             //需要拦截短信,拦截广播
24             abortBroadcast();
25          }
26       }
27    }
28 }
```

代码说明:
- 第 7～10 行的 if 语句用于判断黑名单拦截功能是否开启;
- 第 13 行代码用于获取接收到的短信信息;
- 第 14～20 行代码获取短信中的内容以及电话号码,并对+86 开头的号码进行截取;
- 第 21～25 行代码通过号码查询数据库为哪一种拦截模式,如果是短信拦截或者全部拦截,则直接将该条短信进行拦截。

为了让广播接收者能够拦截短信,必须在 AndroidManifest.xml 文件中注册,具体代码如下:

```
<!-- 拦截黑名单信息 -->
<receiver
android:name="cn.itcast.mobliesafe.chapter03.reciever. InterceptSmsReciever" >
    <intent-filter android:priority="2147483647" >
        <action android:name="android.provider.Telephony.SMS_RECEIVED"/>
    </intent-filter>
</receiver>
```

上述清单文件中,定义了接收短信的广播,并将广播的优先级设置为最高,这样当有新短信到来时会优先被该广播接收者所接收。

3.5.2 拦截电话

当电话铃响时需要自动挂断电话并且不让该记录显示在界面上,而 Google 工程师为了手机的安全性隐藏了挂断电话的服务方法,因此要实现挂断电话的操作只能通过反射获取底层服务。接下来创建拦截电话的广播接收者,具体代码如【文件 3-18】所示。

【文件 3-18】InterceptCallReceiver.java

```
1 public class InterceptCallReciever extends BroadcastReceiver{
2    @Override
3    public void onReceive(Context context,Intent intent){
4       SharedPreferences mSP=context.getSharedPreferences("config",
```

```java
5       Context.MODE_PRIVATE);
6       boolean BlackNumStatus=mSP.getBoolean("BlackNumStatus",true);
7       if(!BlackNumStatus){
8           //黑名单拦截关闭
9           return;
10      }
11      BlackNumberDao dao=new BlackNumberDao(context);
12      if(!intent.getAction().equals(Intent.ACTION_NEW_OUTGOING_CALL)){
13          String mIncomingNumber="";
14          //如果是来电
15          TelephonyManager tManager=(TelephonyManager) context.
16          getSystemService(Service.TELEPHONY_SERVICE);
17          switch(tManager.getCallState()){
18              case TelephonyManager.CALL_STATE_RINGING:
19                  mIncomingNumber=intent.getStringExtra("incoming_number");
20                  int blackContactMode=dao.getBlackContactMode
21                  (mIncomingNumber);
22                  if(blackContactMode==1||blackContactMode==3) {
23                      //观察(另外一个应用程序数据库的变化)呼叫记录的变化,
24                      //如果呼叫记录生成了,就把呼叫记录给删除掉
25                      Uri uri=Uri.parse("content://call_log/calls");
26                      context.getContentResolver().registerContentObserver(
27                      uri,true,new CallLogObserver(new Handler(),
28                      mIncomingNumber, context));
29                      endCall(context);
30                  }
31                  break;
32          }
33      }
34  }
35  /**
36   * 通过内容观察者观察数据库变化
37   */
38  private class CallLogObserver extends ContentObserver {
39      private String incomingNumber;
40      private Context context;
41      public CallLogObserver(Handler handler,String incomingNumber,
42      Context context){
43          super(handler);
```

```java
44          this.incomingNumber=incomingNumber;
45          this.context=context;
46      }
47      //观察到数据库内容变化调用的方法
48      @Override
49      public void onChange(boolean selfChange){
50          Log.i("CallLogObserver","呼叫记录数据库的内容变化了。");
51          context.getContentResolver().unregisterContentObserver(this);
52          deleteCallLog(incomingNumber, context);
53          super.onChange(selfChange);
54      }
55  }
56  /**
57   * 清除呼叫记录
58   * @param incomingNumber
59   */
60  public void deleteCallLog(String incomingNumber,Context context){
61      ContentResolver resolver=context.getContentResolver();
62      Uri uri=Uri.parse("content://call_log/calls");
63      Cursor cursor=resolver.query(uri, new String[] { "_id" }, "number=?",
64          new String[] { incomingNumber }, "_id desc limit 1");
65      if(cursor.moveToNext()){
66          String id=cursor.getString(0);
67          resolver.delete(uri,"_id=?",new String[] {id});
68      }
69  }
70  /**
71   * 挂断电话,需要复制两个AIDL
72   */
73  public void endCall(Context context){
74      try{
75          Class clazz=context.getClassLoader().loadClass
76              ("android.os.ServiceManager");
77          Method method=clazz.getDeclaredMethod("getService",String.class);
78          IBinder iBinder=(IBinder) method.invoke(null,Context.TELEPHONY_SERVICE);
79          ITelephony itelephony=ITelephony.Stub.asInterface(iBinder);
80          itelephony.endCall();
81      } catch(Exception e){
82          e.printStackTrace();
```

```
83        }
84    }
85 }
```

代码说明：

- 第 38~55 行的 CallLogObserver 是一个内容观察者，用于观察系统联系人的数据库，如果黑名单中的电话呼入时，在系统联系人数据库中产生记录时就调用 deleteCallLog() 方法清除历史记录。
- 第 60~69 行的 deleteCallLog() 方法用于清除历史记录，当黑名单中的电话呼入时，手机系统通话记录中会显示该条记录，因此需要把通话记录中的黑名单通话记录删除。手机上拨打电话、接听电话等产生的记录都在系统联系人应用下的 contacts2.db 数据库中，使用 ContentResolver 对象查询并删除数据库中黑名单号码所产生的记录即可。
- 第 73~84 行的 endCall() 方法用于挂断黑名单的呼入电话，该段代码中首先通过反射获取到 ServiceManager 字节码，然后通过该字节码获取 getService() 方法，该方法接收一个 String 类型的参数，然后通过 invoke() 执行 getService() 方法。由于 getService() 方法是静态的，因此 invoke() 的第一个参数可以为 null，第二个参数是 TELEPHONY_SERVICE。由于 getService() 方法的返回值是一个 IBinder 对象（远程服务的代理类），因此需要使用 AIDL 的规则将其转化为接口类型，由于操作是挂断电话，因此需要使用与电话相关的 ITelephony.aidl，然后调用接口中的 endCall() 方法将电话挂断即可。

需要注意的是，与电话相关的操作一般都使用 TelephonyManager 类，但是由于挂断电话的方法在 ITelephony 接口中，而这个接口是隐藏的（@hide），在开发时看不到，因此需要使用 ITelephony.aidl。在使用 ITelephony.aidl 时，需要创建一个与其包名一致的包 com.android. internal. telephony，然后把系统的 ITelephony.aidl 文件复制进来。同时，由于 ITelephony.aidl 接口关联了 NeighboringCellInfo.aidl，也需要一并复制进来。不过要注意的是，NeighboringCellInfo.aidl 所在的包名是 android.telephony，因此需要新建一个 android.telephony 包，然后把 Neighboring CellInfo.aidl 放到该包中。

接下来在 AndroidManifest.xml 文件中注册拦截电话的广播接收者，并且添加与电话有关的权限，具体代码如下：

```
<uses-permission android:name="android.permission.CALL_PHONE"/>
<uses-permission android:name="android.permission.READ_CALL_LOG"/>
<uses-permission android:name="android.permission.WRITE_CALL_LOG"/>
<application
    android:allowBackup="true"
    android:icon="@drawable/ic_launcher"
    android:label="@string/app_name"
    android:theme="@style/AppTheme" >
    ...
    <!-- 拦截黑名单电话 -->
    <receiver android:name="cn.itcast.mobliesafe.chapter03.reciever.
    InterceptCallReciever" >
```

```xml
            <intent-filter android:priority="2147483647" >
                <action android:name="android.intent.action.PHONE_STATE" />
                <action android:name="android.intent.action.NEW_OUTGOING_CALL" />
            </intent-filter>
        </receiver>
</application>
```

至此，通讯卫士模块已经全部完成，接下来对该模块的功能进行完整测试。首先，在添加黑名单界面手动输入或者从联系人列表中选择一个联系人，例如张三，电话号码13070030000，然后使用张三的手机拨打电话，并在电话和短信广播的代码中打上 Log 以便观察结果，这时会看到在 Log 中打印出了电话号码及短信内容，但是手机界面上没有任何显示，说明黑名单拦截功能运行正常。

本章小结

本章主要针对通讯卫士模块进行讲解，首先针对该模块功能进行介绍，然后从创建数据库、创建 UI、编写主界面逻辑代码、编写添加黑名单界面逻辑代码到拦截电话和短信进行了详细讲解。该模块功能复杂，尤其是挂断电话操作使用了 AIDL 进程间通信，编程者要加强学习，并要动手完成所有代码，为学习后面的模块奠定基础。

【面试精选】
1. 请问 Handler 消息机制的原理是什么？应用场景呢？
2. 请问如何将 SQLite 数据库文件与 APK 文件一起发布？
扫描右方二维码，查看面试题答案！

第 4 章 软件管家模块

学习目标

- 了解软件管家模块功能；
- 掌握软件管家模块的开发；
- 了解 dp 与 px 单位互转的工具类。

在日常生活中，每个手机都装有很多程序。为了方便管理这些程序，我们开发了软件管家模块，该模块用于管理手机中的所有程序，当选中某个程序时，可以通过弹出的条目对程序进行启动、卸载、分享和设置。本章将针对软件管家模块进行详细讲解。

4.1 模块概述

4.1.1 功能介绍

软件管家模块用于获取当前手机中的所有应用，以及手机剩余内存、SD 卡剩余内存，并且当点击某个应用时，下方会浮出一个操作条，可以管理程序的启动、卸载、分享、设置。该功能界面效果如图 4-1 所示。

图 4-1 软件管家界面

4.1.2 代码结构

软件管家模块的代码量少、逻辑简单，学起来相对比较容易。接下来通过一个图例来展示软件管家模块的代码结构，如图 4-2 所示。

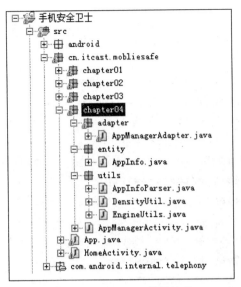

图 4-2 chapter04 代码结构

下面按照结构顺序依次介绍 chapter04 包中的文件，具体如下：
- AppManagerAdapter.java：主界面的数据适配器；
- AppInfo.java：应用程序的实体类，包含程序图标、程序名称、程序大小等字段；
- AppInfoParser.java：用于获取手机中的所有应用程序；
- DensityUtil.java：单位 dip 与 px 互转的工具类；
- EngineUtils.java：启动、卸载、分享、设置应用的业务工具类；
- AppManagerActivity.java：用于展示软件管家界面，主要包括应用列表、手机剩余内存、SD 卡剩余内存。

4.2 软件管家界面

4.2.1 软件管家 UI

软件管家界面主要使用了一个相对布局及一个帧布局，相对布局中放置两个 TextView，分别显示手机剩余内存及 SD 卡剩余内存信息。帧布局中放置一个 ListView 用于展示应用程序，ListView 的上方有一个 TextView 用于说明当前程序是系统程序还是用户程序。需要注意的是，这三个 TextView 中的文本信息都是在代码中动态设置的。软件管家的图形化界面如图 4-3 所示。

图 4-3 所示软件管家界面对应的布局文件如【文件 4-1】所示。

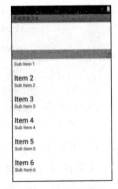

图 4-3 软件管家界面

【文件 4-1】activity_app_manager.xml

```xml
<?xml version="1.0" encoding="utf-8"?>
<LinearLayout xmlns:android="http://schemas.android.com/apk/res/android"
    android:layout_width="match_parent"
    android:layout_height="match_parent"
```

```xml
        android:orientation="vertical" >
        <include layout="@layout/titlebar"/>
        <RelativeLayout
            android:layout_width="match_parent"
            android:layout_height="wrap_content"
            android:layout_margin="5dp" >
            <TextView
                android:id="@+id/tv_phonememory_appmanager"
                style="@style/textview12sp"
                android:layout_alignParentLeft="true"/>
            <TextView
                android:id="@+id/tv_sdmemory_appmanager"
                style="@style/textview12sp"
                android:layout_alignParentRight="true"/>
        </RelativeLayout>
        <FrameLayout
            android:layout_width="match_parent"
            android:layout_height="match_parent" >
            <ListView
                android:id="@+id/lv_appmanager"
                android:layout_width="match_parent"
                android:layout_height="match_parent"
                android:listSelector="@color/transparent"/>
            <TextView
                android:id="@+id/tv_appnumber"
                android:layout_width="match_parent"
                android:layout_height="wrap_content"
                android:background="@color/graye5"
                android:padding="5dp"
                android:textColor="@color/black"/>
        </FrameLayout>
    </LinearLayout>
```

上述布局文件中，有两个 TextView 控件使用了 textview12sp 样式，用于控制字体大小及垂直居中显示等，具体代码如【文件 4-2】所示。

【文件 4-2】res/values/styles.xml

```xml
<!-- TextView 12sp 样式 -->
<style name="textview12sp" parent="wrapcontent">
    <item name="android:textSize">12sp</item>
    <item name="android:gravity">center_vertical</item>
</style>
```

4.2.2 软件管家 Item 布局

当选中某一个应用程序时，程序的下方会浮出一个小条目，提供启动、卸载、分享、设置四个选项可以对程序进行操作，该条目的布局如图 4-4 所示。

图 4-4 中 Item 对应的布局文件如【文件 4-3】所示。

【文件 4-3】item_appmanager_list.xml

图 4-4 Item 布局

```xml
<?xml version="1.0" encoding="utf-8"?>
<LinearLayout xmlns:android="http://schemas.android.com/apk/res/android"
    android:layout_width="match_parent"
    android:layout_height="wrap_content"
    android:orientation="vertical">
    <RelativeLayout
        android:layout_width="match_parent"
        android:layout_height="wrap_content"
        android:layout_margin="5dp">
        <ImageView
            android:id="@+id/imgv_appicon"
            android:layout_width="50dp"
            android:layout_height="50dp"/>
        <TextView
            android:id="@+id/tv_appsize"
            style="@style/textview16sp"
            android:layout_marginRight="10dp"
            android:layout_alignParentRight="true"
            android:layout_centerVertical="true"/>
        <TextView
            android:id="@+id/tv_appname"
            style="@style/textview16sp"
            android:layout_toRightOf="@+id/imgv_appicon"
            android:layout_marginLeft="10dp"
            android:layout_marginTop="5dp"/>
        <TextView
            android:id="@+id/tv_appisroom"
            style="@style/textview14sp"
            android:layout_marginTop="3dp"
            android:layout_alignLeft="@+id/tv_appname"
            android:layout_toRightOf="@+id/imgv_appicon"
```

```xml
            android:layout_below="@+id/tv_appname"/>
    </RelativeLayout>
    <View
        android:layout_width="match_parent"
        android:layout_height="1px"
        android:background="@color/black30"/>
    <LinearLayout
        android:id="@+id/ll_option_app"
        android:layout_width="match_parent"
        android:layout_height="wrap_content"
        android:orientation="horizontal"
        android:layout_marginTop="3dp"
        android:visibility="visible">
        <TextView
            android:id="@+id/tv_launch_app"
            style="@style/tvyellow14sp"
            android:drawableTop="@drawable/phoneicon_yellow_n"
            android:text="启动"/>
        <View
            android:layout_width="1px"
            android:layout_height="match_parent"
            android:background="@color/black30"/>
        <TextView
            android:id="@+id/tv_uninstall_app"
            style="@style/tvyellow14sp"
            android:drawableTop="@drawable/deleteicon_yellow_n"
            android:text="卸载"/>
        <View
            android:layout_width="1px"
            android:layout_height="match_parent"
            android:background="@color/black30"/>
        <TextView
            android:id="@+id/tv_share_app"
            style="@style/tvyellow14sp"
            android:text="分享"
            android:drawableTop="@drawable/shareicon_yellow_n"/>
        <View
            android:layout_width="1px"
            android:layout_height="match_parent"
```

```xml
            android:background="@color/black30"/>
        <TextView
            android:id="@+id/tv_setting_app"
            style="@style/tvyellow14sp"
            android:text="设置"
            android:drawableTop="@drawable/settingicon_yellow_n"/>
    </LinearLayout>
</LinearLayout>
```

上述布局文件中，主要使用了一个相对布局和一个线性布局，相对布局中放置了一个 ImageView 控件和三个 TextView 控件，分别用于展示应用程序的图标，以及应用名称、存储在手机还是 SD 卡中、该应用程序在手机中占用的内存量；线性布局中放置了四个 TextView 控件和三个 View 控件，分别用于展示这四个按钮的文字提示、功能图标以及功能图标左侧分隔线。

软件管家的应用列表布局以及弹出的小条目布局都使用了 TextView 控件，为了提高代码的复用性，可以将文本样式定义在 styles.xml 文件中，在布局中进行引用，具体代码如【文件 4-4】所示。

【文件 4-4】res/values/styles.xml

```xml
<resources>
    <!-- TextView 14sp 样式 -->
    <style name="textview14sp" parent="wrapcontent">
        <item name="android:textSize">14sp</item>
        <item name="android:gravity">center_vertical</item>
    </style>
    <!-- 宽度为 0dp，高度为包裹内容，权重为 1 -->
    <style name="zero_widthwrapcontent">
        <item name="android:layout_width">0dp</item>
        <item name="android:layout_height">wrap_content</item>
        <item name="android:layout_weight">1</item>
    </style>
    <!-- 黄色条目中的 TextView 控件，字体 14sp，图片间距为 5dp 的 textview -->
    <style name="tvyellow14sp" parent="zero_widthwrapcontent">
        <item name="android:textSize">14sp</item>
        <item name="android:drawablePadding">3dp</item>
        <item name="android:gravity">center</item>
        <item name="android:textColor">@color/bright_yellow</item>
    </style>
</resources>
```

上述代码自定义了三个样式，其中最后一个 tvyellow14sp 样式继承了 zero_widthwrapcontent 样式，表示同时拥有这两个自定义样式的所有属性。

4.2.3 应用程序的实体类

在获取应用程序列表之前首先需要创建一个实体类（AppInfo.java），该类用于存储应用程序的相关信息，如程序包名、程序图标、程序大小等，具体代码如【文件 4-5】所示。

【文件 4-5】 AppInfo.java

```java
public class AppInfo {
    /** 应用程序包名 */
    public String packageName;
    /** 应用程序图标 */
    public Drawable icon;
    /** 应用程序名称 */
    public String appName;
    /** 应用程序路径 */
    public String apkPath;
    /** 应用程序大小 */
    public long appSize;
    /** 是否是手机存储 */
    public boolean isInRoom;
    /** 是否是用户应用 */
    public boolean isUserApp;
    /** 是否选中，默认都为 false */
    public boolean isSelected = false;
    /** 拿到 App 位置字符串 */
    public String getAppLocation(boolean isInRoom) {
        if(isInRoom){
            return "手机内存";
        }else{
            return "外部存储";
        }
    }
}
```

上述代码中，定义了一个 getAppLocation() 方法，该方法用于返回一个字符串，表示当前 APP 是存储在手机内存中还是存储在外部设备上。

4.3 工具类

4.3.1 获取应用程序信息

在开发软件管家模块时，需要获取手机中所有已安装程序，由于该功能比较独立，因此，可以将其专门定义一个类，以提高代码的可复用性，具体代码如【文件 4-6】所示。

【文件 4-6】 AppInfoParser.java

```java
public class AppInfoParser{
    /**
     * 获取手机里面的所有的应用程序
     * @param context 上下文
     * @return
     */
    public static List<AppInfo> getAppInfos(Context context){
        //得到一个包管理器
        PackageManager pm=context.getPackageManager();
        List<PackageInfo> packInfos=pm.getInstalledPackages(0);
        List<AppInfo> appinfos=new ArrayList<AppInfo>();
        for(PackageInfo packInfo:packInfos){
            AppInfo appinfo=new AppInfo();
            String packname=packInfo.packageName;
            appinfo.packageName=packname;
            Drawable icon=packInfo.applicationInfo.loadIcon(pm);
            appinfo.icon=icon;
            String appname=packInfo.applicationInfo.loadLabel(pm).toString();
            appinfo.appName=appname;
            //应用程序apk包的路径
            String apkpath=packInfo.applicationInfo.sourceDir;
            appinfo.apkPath=apkpath;
            File file=new File(apkpath);
            long appSize=file.length();
            appinfo.appSize=appSize;
            //应用程序安装的位置
            int flags=packInfo.applicationInfo.flags; //二进制映射
            if ((ApplicationInfo.FLAG_EXTERNAL_STORAGE & flags)!=0){
                //外部存储
                appinfo.isInRoom=false;
            } else {
                //手机内存
                appinfo.isInRoom=true;
            }
            if((ApplicationInfo.FLAG_SYSTEM & flags)!=0){
                //系统应用
                appinfo.isUserApp=false;
            } else {
```

```
39              //用户应用
40              appinfo.isUserApp=true;
41          }
42          appinfos.add(appinfo);
43          appinfo = null;
44      }
45      return appinfos;
46  }
47 }
```

代码说明:

- 第 9~10 行代码用于获取包管理器,通过包管理器获取手机中已安装的应用程序的包信息,并存储到 List<PackageInfo>集合中。
- 第 11 行代码定义了一个 List<AppInfo>集合,该集合用于存储获取到的应用程序信息。
- 第 12~19 行代码用于遍历 packInfos 集合对象,并将每个应用程序的包名、图标、应用名称存储到 AppInfo 对象中。
- 第 21~25 行代码用于获取应用程序的大小,首先获取到应用程序的路径,然后通过 File 对象获取到应用程序的大小,并存储到 AppInfo 对象中。
- 第 27 行代码用于获取程序的二进制映射的 flags。
- 第 28~34 行代码用于判断程序的安装位置,通过第 28 行获取到的 flags 判断应用程序是否安装在外部存储设备中,如果是,则将 AppInfo 的 isInRoom 属性设置为 false,否则将 isInRoom 属性设置为 true。
- 第 35~41 行代码通过 flags 判断程序是否为系统应用,如果判断条件 (ApplicationInfo.FLAG_SYSTEM&flags)!=0 成立,则是系统应用,将 AppInfo 的 isUerApp 属性设置为 false,否则就是用户应用,将 isUserApp 属性设置为 true。
- 第 42~43 行代码将 AppInfo 对象添加到 List<AppInfo>集合中,然后将 AppInfo 对象设置为 null,将之前获取的应用程序的信息删除,用于存储再次循环得到的应用程序信息。

4.3.2 单位转换

在布局文件中设置单位时,既可以用 px,也可以用 dip(或者 dp),但通常情况下会使用 dip,这样可以保证在不同屏幕分辨率的机器上布局一致。但是在编写代码时,有很多控件中都只提供了设置 px 的方法,例如 AppManagerAdapter 文件中动态设置 TextView 的 setPadding() 方法并没有提供设置 dip 的方法,此时,就需根据手机的分辨率将单位从 dip 转换成 px。接下来定义一个工具类 DensityUtil,方便 dip 与 px 之间的转换,具体代码如【文件 4-7】所示。

【文件 4-7】DensityUtil.java

```
1 public class DensityUtil{
2    /**
3     * dip 转换像素 px
4     */
```

```
5   public static int dip2px(Context context,float dpValue){
6       try{
7           final float scale=context.getResources().getDisplayMetrics().density;
8           return (int) (dpValue*scale+0.5f);
9       } catch(Exception e){
10          e.printStackTrace();
11      }
12      return (int) dpValue;
13  }
14  /**
15   * 像素 px 转换为 dip
16   */
17  public static int px2dip(Context context,float pxValue){
18      try {
19          final float scale=context.getResources().getDisplayMetrics().density;
20          return (int) (pxValue/scale+0.5f);
21      } catch (Exception e) {
22          e.printStackTrace();
23      }
24      return (int) pxValue;
25  }
26  }
```

上述代码进行单位转换时，首先通过 context.getResources().getDisplayMetrics().density 获取屏幕的分辨率，然后使用公式 dpValue*scale+0.5f 计算 dip 转换为 px 的值，使用公式 pxValue/ scale+0.5f 计算 px 转换为 dip 的值（该工具类会用即可，可直接从网上下载）。

4.3.3 程序的业务类

当用户选中某个条目时，隐藏在下面的线性布局就会弹出，展示"启动""卸载""分享""设置"四个选项。为了让代码逻辑更加清晰，现将这四个功能抽取出来作为一个工具类，具体代码如【文件 4-8】所示。

【文件 4-8】 EngineUtils.java

```
1  /** 业务工具类 */
2  public class EngineUtils{
3      /**
4       * 分享应用
5       */
6      public static void shareApplication(Context context, AppInfo appInfo){
7          Intent intent=new Intent("android.intent.action.SEND");
8          intent.addCategory("android.intent.category.DEFAULT");
```

```java
9          intent.setType("text/plain");
10         intent.putExtra(Intent.EXTRA_TEXT, "推荐您使用一款软件,名称叫: "+
11         appInfo.appName+
12         "下载路径: https://play.google.com/store/apps/details?id="+
13         appInfo.packageName);
14         context.startActivity(intent);
15     }
16     /**
17      * 开启应用程序
18      */
19     public static void startApplication(Context context,AppInfo appInfo){
20         //打开这个应用程序的入口activity
21         PackageManager pm=context.getPackageManager();
22         Intent intent=pm.getLaunchIntentForPackage(appInfo.packageName);
23         if(intent!=null){
24             context.startActivity(intent);
25         }else{
26             Toast.makeText(context,"该应用没有启动界面",0).show();
27         }
28     }
29     /**
30      * 开启应用设置页面
31      * @param context
32      * @param appInfo
33      */
34     public static void settingAppDetail(Context context, AppInfo appInfo){
35         Intent intent=new Intent();
36         intent.setAction("android.settings.APPLICATION_DETAILS_SETTINGS");
37         intent.addCategory(Intent.CATEGORY_DEFAULT);
38         intent.setData(Uri.parse("package:"+appInfo.packageName));
39         context.startActivity(intent);
40     }
41     /** 卸载应用 */
42     public static void uninstallApplication(Context context, AppInfo appInfo){
43         if(appInfo.isUserApp){
44             Intent intent=new Intent();
45             intent.setAction(Intent.ACTION_DELETE);
46             intent.setData(Uri.parse("package:"+appInfo.packageName));
47             context.startActivity(intent);
```

```
48        } else {
49            //系统应用需要root权限,利用linux命令删除文件
50            if(!RootTools.isRootAvailable()){
51                Toast.makeText(context,"卸载系统应用,必须要root权限",0).show();
52                return;
53            }
54            try{
55                if(!RootTools.isAccessGiven()){
56                    Toast.makeText(context,"请授权黑马小护卫root权限",0).show();
57                    return;
58                }
59                RootTools.sendShell("mount -o remount ,rw /system",3000);
60                RootTools.sendShell("rm -r "+appInfo.apkPath,30000);
61            } catch(Exception e){
62                e.printStackTrace();
63            }
64        }
65    }
66 }
```

代码说明：

- 第6~15行的shareApplication()方法用于分享应用程序，通过隐式意图激活发送短信功能，然后指定Category为默认值，Type为text/plain，并通过putExtra属性将文本加入到发送的短信中。
- 第19~28行的startApplication()方法用于开启当前应用，首先获取PackageManager包管理器，然后通过getLaunchIntentForPackage()方法获取一个安装包启动的意图，将包名传入该方法，通过意图即可启动程序。
- 第34~40行的SettingAppDetail()方法用于开启应用程序设置界面，首先将Action设置为android.settings.APPLICATION_DETAILS_SETTINGS，Category为默认值，将Data设置为应用程序的包名，然后开启隐式意图。
- 第42~65行的uninstallApplication()方法用于卸载应用程序，其中第43~48行代码判断程序是否为用户程序，如果是则通过隐式意图直接删除。如果是系统应用，需要Root权限才能卸载。在Android系统中获取Root权限需要两步，首先通过RootTools.isRootAvailable()方法判断是否有Root权限，如果无，则通过RootTools.isAccessGiven()方法获取Root权限。获取到权限后，通过Linux命令即可删除应用。

4.4 软件管家功能

4.4.1 软件管家逻辑

所有准备工作都已做完，接下来编写软件管家界面的逻辑代码，将手机内存信息以及应

用程序列表等数据加载到界面中，具体代码如【文件 4-9】所示。

【文件 4-9】 AppManagerActivity.java

```java
1  public class AppManagerActivity extends Activity implements OnClick Listener{
2      /** 手机剩余内存 TextView */
3      private TextView mPhoneMemoryTV;
4      /** 展示 SD 卡剩余内存 TextView */
5      private TextView mSDMemoryTV;
6      private ListView mListView;
7      private List<AppInfo> appInfos;
8      private List<AppInfo> userAppInfos=new ArrayList<AppInfo>();
9      private List<AppInfo> systemAppInfos=new ArrayList<AppInfo>();
10     private AppManagerAdapter adapter;
11     /** 接收应用程序卸载成功的广播 */
12     private UninstallRececiver receciver;
13     private Handler mHandler=new Handler(){
14         public void handleMessage(android.os.Message msg){
15             switch(msg.what){
16             case 10:
17                 if(adapter==null){
18                     adapter=new AppManagerAdapter(userAppInfos,
19                         systemAppInfos, AppManagerActivity.this);
20                 }
21                 mListView.setAdapter(adapter);
22                 adapter.notifyDataSetChanged();
23                 break;
24             case 15:
25                 adapter.notifyDataSetChanged();
26                 break;
27             }
28         };
29     };
30     private TextView mAppNumTV;
31     @Override
32     protected void onCreate(Bundle savedInstanceState){
33         super.onCreate(savedInstanceState);
34         requestWindowFeature(Window.FEATURE_NO_TITLE);
35         setContentView(R.layout.activity_app_manager);
36         //注册广播
37         receciver=new UninstallRececiver();
```

```
38      IntentFilter intentFilter=new IntentFilter(Intent.ACTION_PACKAGE_REMOVED);
39      intentFilter.addDataScheme("package");
40      registerReceiver(receciver,intentFilter);
41      initView();
42  }
43  /** 初始化控件 */
44  private void initView(){
45      findViewById(R.id.rl_titlebar).setBackgroundColor(
46      getResources().getColor(R.color.bright_yellow));
47      ImageView mLeftImgv=(ImageView) findViewById(R.id.imgv_leftbtn);
48      ((TextView) findViewById(R.id.tv_title)).setText("软件管家");
49      mLeftImgv.setOnClickListener(this);
50      mLeftImgv.setImageResource(R.drawable.back);
51      mPhoneMemoryTV=(TextView) findViewById(R.id.tv_phonememory_appmanager);
52      mSDMemoryTV=(TextView) findViewById(R.id.tv_sdmemory_appmanager);
53      mAppNumTV=(TextView) findViewById(R.id.tv_appnumber);
54      mListView=(ListView) findViewById(R.id.lv_appmanager);
55      //取得手机剩余内存和SD卡剩余内存
56      getMemoryFromPhone();
57      initData();
58      initListener();
59  }
60  private void initListener(){
61      mListView.setOnItemClickListener(new OnItemClickListener(){
62          @Override
63          public void onItemClick(AdapterView<?> parent,View view,
64          final int position,long id){
65              if(adapter!=null){
66                  new Thread(){
67                      public void run(){
68                          AppInfo mappInfo=(AppInfo) adapter.getItem(position);
69                          //记住当前条目的状态
70                          boolean flag=mappInfo.isSelected;
71                          //先将集合中所有条目的AppInfo变为未选中状态
72                          for(AppInfo appInfo:userAppInfos){
73                              appInfo.isSelected=false;
74                          }
75                          for(AppInfo appInfo:systemAppInfos){
76                              appInfo.isSelected=false;
```

```
77                          }
78                      if(mappInfo!=null){
79                          //如果已经选中，则变为未选中
80                          if(flag){
81                              mappInfo.isSelected=false;
82                          }else{
83                              mappInfo.isSelected=true;
84                          }
85                          mHandler.sendEmptyMessage(15);
86                      }
87                  };
88              }.start();
89          }
90      }
91  });
92  mListView.setOnScrollListener(new OnScrollListener(){
93      @Override
94      public void onScrollStateChanged(AbsListView view, int scrollState) {
95      }
96      @Override
97      public void onScroll(AbsListView view,int firstVisibleItem,
98          int visibleItemCount,int totalItemCount){
99          if(firstVisibleItem>=userAppInfos.size()+1){
100             mAppNumTV.setText("系统程序: "+systemAppInfos.size()+"个");
101         } else{
102             mAppNumTV.setText("用户程序: " + userAppInfos.size() + "个");
103         }
104     }
105 });
106 }
107 private void initData(){
108     appInfos=new ArrayList<AppInfo>();
109     new Thread(){
110         public void run(){
111             appInfos.clear();
112             userAppInfos.clear();
113             systemAppInfos.clear();
114             appInfos.addAll(AppInfoParser.getAppInfos(AppManager
115                 Activity.this));
```

```java
116             for(AppInfo appInfo:appInfos){
117                 //如果是用户 App
118                 if(appInfo.isUserApp){
119                     userAppInfos.add(appInfo);
120                 }else{
121                     systemAppInfos.add(appInfo);
122                 }
123             }
124             mHandler.sendEmptyMessage(10);
125         };
126     }.start();
127 }
128 /** 拿到手机和 SD 卡剩余内存 */
129 private void getMemoryFromPhone(){
130     long avail_sd=Environment.getExternalStorageDirectory().getFreeSpace();
131     long avail_rom=Environment.getDataDirectory().getFreeSpace();
132     //格式化内存
133     String str_avail_sd=Formatter.formatFileSize(this,avail_sd);
134     String str_avail_rom=Formatter.formatFileSize(this,avail_rom);
135     mPhoneMemoryTV.setText("剩余手机内存: "+str_avail_rom);
136     mSDMemoryTV.setText("剩余SD卡内存: "+str_avail_sd);
137 }
138 @Override
139 public void onClick(View v){
140     switch(v.getId()){
141     case R.id.imgv_leftbtn:
142         finish();
143         break;
144     }
145 }
146 /**
147  * 接收应用程序卸载的广播
148  * @author admin
149  */
150 class UninstallRececiver extends BroadcastReceiver{
151     @Override
152     public void onReceive(Context context,Intent intent){
153         //收到广播了
154         initData();
```

```
155        }
156    }
157    @Override
158    protected void onDestroy(){
159        unregisterReceiver(receciver);
160        receciver=null;
161        super.onDestroy();
162    }
163 }
```

- 第 13~29 行代码定义了一个 Handler，用于接收发送的消息，当接收的消息为 10 时，则使用 if 语句判断 Adapter 是否为空，如果为空，则创建 AppManagerAdapter 对象，并将数据填充给 ListView 控件，然后调用 adapter.notifyDataSetChanged()更新 ListView 界面。当接收到的消息为 15 时，则直接调用 notifyDataSetChanged()更新 ListView 界面。
- 第 37~40 行代码通过代码动态的注册了一个广播接收者，用于接收应用卸载成功的广播。
- 第 44~59 行的 initView()方法主要用于初始化控件，然后分别调用 getMemoryFromPhone()方法、initData()方法、initLinstener()方法初始化数据。
- 第 60~106 行的 initListener()方法为 ListView 注册了一个条目点击事件，首先获取当前点击条目对应的 AppInfo 对象，并通过 boolean flag = mappInfo.isSelected 记录当前的状态是否被选中，然后将集合中的用户程序和系统程序均设置未被选中状态。最后判断当前 AppInfo 对象是否为 null，不为 null 则判断是否为未选中状态，如果是则设置 mappInfo.isSelected = false;，否则 mappInfo.isSelected = true;，并发送 Message 给 Handler 对象。其中 92~105 行代码为 ListView 注册了一个条目滚动事件，并在 onScroll()方法中判断当前程序是用户程序还是系统程序，如果是系统程序，则将界面中的 TextView 标签上的文字动态设置为"系统程序：+个数"，否则设置为"用户程序：+个数"。
- 第 107~127 行的 initData()方法用于初始化数据，获取到系统中的所有程序，并按照用户程序和系统程序分别添加到 userAppInfos 和 systemAppInfos 集合中，最后通过 mHandler.sendEmptyMessage(10)发送 Message 给 Handler 对象更新界面。
- 第 129~137 行的 getMemoryFromPhone()方法用于获取手机本身和 SD 卡剩余内存，由于这个值的类型为 long 比较长，因此使用 Formatter.formatFileSize()方法进行格式化，返回一个 string 类型的值，单位为 KB 或 MB。
- 第 139~145 行的 onClick()方法当点击界面左上角的返回按钮时,关闭当前界面显示主界面。
- 第 150~156 行代码用于定义一个广播接收者，用于接收程序卸载的广播，当程序卸载完成后，调用 initData()方法，清除数据重新加载应用列表，更新界面。

4.4.2 数据适配器

由于软件管家界面中使用了 ListView 控件，因此需要创建一个数据适配器，将所有应用程序信息填充到界面中，数据适配器代码如【文件 4-10】所示。

【文件 4-10】 AppManagerAdapter.java

```java
public class AppManagerAdapter extends BaseAdapter{
    private List<AppInfo> UserAppInfos;
    private List<AppInfo> SystemAppInfos;
    private Context context;
    public AppManagerAdapter(List<AppInfo> userAppInfos,List<AppInfo>
systemAppInfos,Context context){
        super();
        UserAppInfos=userAppInfos;
        SystemAppInfos=systemAppInfos;
        this.context=context;
    }
    @Override
    public int getCount(){
        //因为有两个条目需要用于显示用户进程，因此系统进程需要加2
        return UserAppInfos.size()+SystemAppInfos.size()+2;
    }
    @Override
    public Object getItem(int position){
        if(position==0){
            //第0个位置显示的应该是用户程序个数的标签
            return null;
        } else if(position==(UserAppInfos.size()+1)){
            return null;
        }
        AppInfo appInfo;
        if(position<(UserAppInfos.size()+1)){
            // 用户程序
            appInfo=UserAppInfos.get(position-1);
            //多了一个textview标签，位置-1
        } else {
            //系统程序
            int location=position-UserAppInfos.size()-2;
            appInfo=SystemAppInfos.get(location);
        }
        return appInfo;
    }
    @Override
    public long getItemId(int position){
```

```java
39      //TODO Auto-generated method stub
40      return position;
41  }
42  @Override
43  public View getView(int position,View convertView,ViewGroup parent){
44      //如果 position 为 0，则为 TextView
45      if(position==0){
46          TextView tv=getTextView();
47          tv.setText("用户程序: "+UserAppInfos.size()+"个");
48          return tv;
49          //系统应用
50      } else if(position==(UserAppInfos.size()+1)){
51          TextView tv=getTextView();
52          tv.setText("系统程序: "+SystemAppInfos.size()+"个");
53          return tv;
54      }
55      //获取到当前App的对象
56      AppInfo appInfo;
57      if(position<(UserAppInfos.size()+1)){
58          //position 0 为 textView
59          appInfo=UserAppInfos.get(position-1);
60      } else {
61          //系统应用
62          appInfo = SystemAppInfos.get(position-UserAppInfos.size()-2);
63      }
64      ViewHolder viewHolder=null;
65      if (convertView!=null & convertView instanceof LinearLayout){
66          viewHolder=(ViewHolder) convertView.getTag();
67      } else {
68          viewHolder=new ViewHolder();
69          convertView=View.inflate(context, R.layout.item_appmanager_list,null);
70          viewHolder.mAppIconImgv=(ImageView) convertView.
71          findViewById(R.id.imgv_appicon);
72          viewHolder.mAppLocationTV=(TextView) convertView.
73          findViewById(R.id.tv_appisroom);
74          viewHolder.mAppSizeTV=(TextView) convertView.
75          findViewById(R.id.tv_appsize);
76          viewHolder.mAppNameTV=(TextView) convertView.
77          findViewById(R.id.tv_appname);
```

```java
78       viewHolder.mLuanchAppTV=(TextView) convertView.
79         findViewById(R.id.tv_launch_app);
80       viewHolder.mSettingAppTV=(TextView) convertView.
81         findViewById(R.id.tv_setting_app);
82       viewHolder.mShareAppTV=(TextView) convertView.
83         findViewById(R.id.tv_share_app);
84       viewHolder.mUninstallTV=(TextView) convertView.
85         findViewById(R.id.tv_uninstall_app);
86       viewHolder.mAppOptionLL=(LinearLayout) convertView.
87         findViewById(R.id.ll_option_app);
88       convertView.setTag(viewHolder);
89     }
90     if(appInfo!=null){
91       viewHolder.mAppLocationTV.setText(appInfo.
92         getAppLocation(appInfo.isInRoom));
93       viewHolder.mAppIconImgv.setImageDrawable(appInfo.icon);
94       viewHolder.mAppSizeTV.setText(Formatter.formatFileSize(context,
95         appInfo.appSize));
96       viewHolder.mAppNameTV.setText(appInfo.appName);
97       if(appInfo.isSelected){
98         viewHolder.mAppOptionLL.setVisibility(View.VISIBLE);
99       }else{
100        viewHolder.mAppOptionLL.setVisibility(View.GONE);
101      }
102    }
103    MyClickListener listener=new MyClickListener(appInfo);
104    viewHolder.mLuanchAppTV.setOnClickListener(listener);
105    viewHolder.mSettingAppTV.setOnClickListener(listener);
106    viewHolder.mShareAppTV.setOnClickListener(listener);
107    viewHolder.mUninstallTV.setOnClickListener(listener);
108    return convertView;
109  }
110
111  /**
112   * 创建一个TextView
113   * @return
114   */
115  private TextView getTextView(){
116    TextView tv=new TextView(context);
```

```java
117        tv.setBackgroundColor(context.getResources().getColor(color.graye5));
118        tv.setPadding(DensityUtil.dip2px(context,5),
119        DensityUtil.dip2px(context,5), DensityUtil.dip2px(context, 5),
120        DensityUtil.dip2px(context,5));
121        tv.setTextColor(context.getResources().getColor(color.black));
122        return tv;
123    }
124    static class ViewHolder{
125        /** 启动 App */
126        TextView mLuanchAppTV;
127        /** 卸载 App */
128        TextView mUninstallTV;
129        /** 分享 App */
130        TextView mShareAppTV;
131        /** 设置 App */
132        TextView mSettingAppTV;
133        /** app 图标 */
134        ImageView mAppIconImgv;
135        /** app 位置 */
136        TextView mAppLocationTV;
137        /** app 大小 */
138        TextView mAppSizeTV;
139        /** app 名称 */
140        TextView mAppNameTV;
141        /** 操作 App 的线性布局 */
142        LinearLayout mAppOptionLL;
143    }
144    class MyClickListener implements OnClickListener{
145        private AppInfo appInfo;
146        public MyClickListener(AppInfo appInfo){
147            super();
148            this.appInfo=appInfo;
149        }
150        @Override
151        public void onClick(View v){
152            switch(v.getId()){
153            case R.id.tv_launch_app:
154                //启动应用
155                EngineUtils.startApplication(context,appInfo);
```

```java
156             break;
157         case R.id.tv_share_app:
158             //分享应用
159             EngineUtils.shareApplication(context,appInfo);
160             break;
161         case R.id.tv_setting_app:
162             //设置应用
163             EngineUtils.settingAppDetail(context,appInfo);
164             break;
165         case R.id.tv_uninstall_app:
166             //卸载应用,需要注册广播接收者
167             if(appInfo.packageName.equals(context.getPackageName())){
168                 Toast.makeText(context,"您没有权限卸载此应用! ",0).show();
169                 return;
170             }
171             EngineUtils.uninstallApplication(context,appInfo);
172             break;
173         }
174     }
175 }
176}
```

代码说明：

- 第5~11行的AppManagerAdapter()方法为该适配器类的构造方法，该方法中接收三个参数，第一个参数是一个用于存储用户程序的List集合，第二个参数是一个用于存储系统程序的List集合，第三个参数是应用程序的上下文。在构造方法中将接收到的List数据赋值给UserAppInfos和SystemAppInfos集合。
- 第13~16行的getCount()方法用于显示当前条目的个数，由于当前条目中添加了两个TextView用于显示系统程序和用户程序，因此个数应该为用户应用的个数加上系统应用的个数再加上2，也就是UserAppInfos.size() + SystemAppInfos.size()+2。
- 第18~36行的getItem()方法用于返回当前位置的条目对象，首先判断当前位置是否为"用户程序"和"系统程序"的TextView标签，是则返回null。接下来判断是否为用户程序，如果是则返回UserAppInfos.get(position-1)位置对应的AppInfo对象，否则就是系统程序，系统程序则返回SystemAppInfos.get(position - UserAppInfos.size() - 2);位置对应的AppInfo对象。
- 第38~41行的getItemId()方法用于返回当前位置的ItemId。
- 第44~53行代码用于判断当前条目是否为"用户程序"和"系统程序"的TextView标签，如果是，则指定文本显示内容为用户程序或系统程序的个数。
- 第56~63行代码用于获取当前AppInfo对象，包括用户程序和系统程序。
- 第64~108行代码通过ViewHolder进行填充数据，并设置某些控件的监听事件。

- 第 115~123 行代码用于动态的创建一个 TextView 控件，用于现在用户程序和系统程序的标签文本。将 TextView 条目背景色设置为灰色，文字颜色设置为白色，并指定文字与父控件的上、下、左、右边界的距离（DensityUtil 是一个单位转换工具类，在 4.3.2 小节已讲解）。
- 第 144~175 行的 MyClickListener 类实现了 OnClickListener 接口，并实现了该接口中的 onClick()方法，在该方法中通过 switch 语句判断点击的是"启动""卸载""分享""设置"，然后调用 EngineUtils 工具类中的相应方法执行不同的操作。

本 章 小 结

本章主要是针对手机卫士中的软件管家模块进行讲解，首先针对该模块功能、代码结构进行介绍，然后讲解了软件管家的 UI 界面，最后实现软件管家中的逻辑代码。通过该模块可以学习到 ListView 的优化、获取应用程序信息、dp 与 px 之间的转换。该模块功能较为简单，很容易理解，编程者只需要按照步骤完成模块的开发即可。

【面试精选】
1. 请问 Android 中的 dip、px 和 sp 单位有何区别？
2. 请问注册广播有几种方式，这几种方式有何不同？
扫描右方二维码，查看面试题答案！

手机杀毒模块

学习目标
- 了解手机杀毒模块功能；
- 掌握第三方数据库的使用；
- 掌握病毒查杀的原理；
- 了解应用程序的 MD5 码。

大家在使用计算机时，最烦恼的事情就是被病毒感染，病毒会干扰计算机系统，并破坏系统中的程序，有些病毒还会窃取用户隐私和资料。同样，在使用手机时也会遇到这种情况，只不过手机中的病毒都存在于 APK 文件中，只要将病毒所在的 APK 文件删除即可将其清理掉。本章将针对手机杀毒模块进行详细讲解。

5.1 模块概述

5.1.1 功能介绍

手机杀毒模块主要用于全盘扫描手机程序，检测是否有病毒。当点击"全盘扫描"图标时，会显示当前的查杀进度以及正在扫描的程序，如果某个程序是病毒，则应用名称会显示成红色，手动卸载程序即可，如果没有病毒则显示扫描完成。当下次再进入手机杀毒模块时，会显示上次病毒查杀的时间。该功能界面效果如图 5-1 所示。

图 5-1　手机杀毒界面

5.1.2 代码结构

手机杀毒模块的核心代码就是查询程序的 MD5 码是否在病毒数据库中，如果在数据库中那么该程序就是病毒，否则就不是病毒。接下来通过一个图例来展示手机杀毒模块的代码结构，如图 5-2 所示。

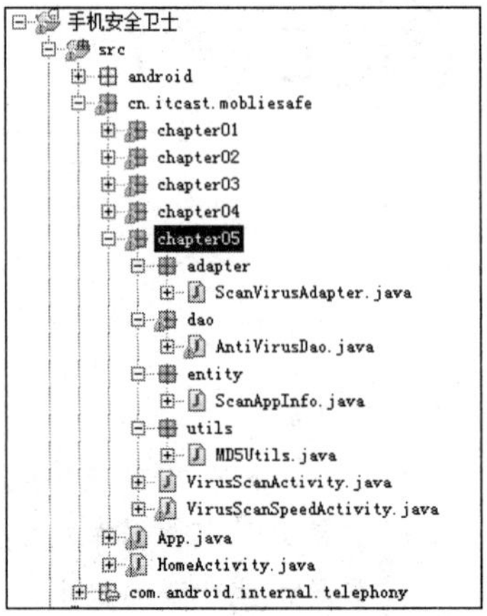

下面按照结构顺序依次介绍 chapter05 包中的文件，具体如下：

- ScanVirusAdapter.java：杀毒界面 ListView 的适配器；
- AntiVirusDao.java：病毒数据库的操作类；
- ScanAppInfo.java：应用程序的实体类；
- MD5Utils.java：获取文件 MD5 码的工具类；
- VirusScanActivity.java：手机杀毒模块的主界面逻辑，主要包含上次查杀时间与跳转按钮；
- VirusScanSpeedActivity.java：手机杀毒模块中查杀病毒逻辑。

图 5-2　chapter05 代码结构

5.1.3 手机病毒

手机病毒是一种具有传染性、破坏性的手机程序，可用杀毒软件进行清除与查杀，也可以手动卸载。其可利用短信、彩信、电子邮件、蓝牙等方式进行传播，会导致用户手机死机、关机、个人资料被删、个人信息泄露、自动拨打电话、发短（彩）信等进行恶意扣费，甚至会损毁 SIM 卡、芯片等硬件，导致手机无法正常使用。

历史上最早的手机病毒出现在 2000 年，当时，手机公司 Movistar 收到大量由计算机发出的名为 Timofonica 的骚扰短信，该病毒通过某国电信公司 Telefonica 的移动系统向系统内的用户发送脏话等垃圾短信。事实上，该病毒最多只能算作短信炸弹。真正意义上的手机病毒是在 2004 年 6 月出现的，那就是 Cabir 蠕虫病毒，这种病毒通过某品牌 S60 系列手机复制，然后不断寻找安装了蓝牙的手机。从此以后，手机病毒开始泛滥。

手机中的软件都安装在系统中，这个系统是一个嵌入式操作系统（固化在芯片中的操作系统，一般由 Java、C++等语言编写），相当于一个小型的智能处理器，手机病毒就是靠系统的漏洞来入侵手机的。手机病毒要传播和运行，必须要有移动服务商提供数据传输功能，而且手机需要支持 Java 等高级程序的写入功能。许多具备上网及下载等功能的手机都可能会被手机病毒入侵。手机病毒一般具有以下几个特点：

- 传染性：病毒通过自身复制来感染正常文件，达到破坏目标程序的目的，但是它的感染是有条件的，也就是病毒程序必须被执行之后才具有传染性，才能感染其他文件。
- 破坏性：任何病毒侵入目标后，都会或大或小地对系统正常使用造成一定的影响，轻者降低系统的性能，占用系统资源，重者破坏数据导致系统崩溃，甚至损坏硬件。
- 潜伏性：一般病毒在感染文件后不是立即发作，而是隐藏在系统中，在满足条件时才能激活。病毒如果没有被激活，它就像其他没执行的程序一样，安静地待在系统中，没有传染性也不具有杀伤力。

- 寄生性：病毒嵌入到载体中，依靠载体而生存，当载体被执行时，病毒程序也就被激活，然后进行复制和传播。

手机病毒的感染主要来源于应用程序的安装，要想杀死某个病毒，只需要卸载病毒程序即可。每一个应用程序都有对应的特征码，这个特征码称为 MD5 码。MD5 码是程序的唯一标识，手机杀毒厂商搜集了大量手机病毒应用的 MD5 码存入数据库中，当杀毒软件对手机应用进行扫描时，会先得到被扫描应用的 MD5 码，并将该 MD5 码与病毒数据库中存储的 MD5 码进行比对，如果发现应用程序的 MD5 码存在于数据库中，则说明这个应用是病毒，可以直接清除或者手动卸载该应用。

要防范手机感染病毒，最好在安装程序时就要做好防范，例如不要点击手机短信中的 APK 链接下载应用；不要扫描不安全的二维码下载应用。由于有些病毒会伪装成市面上较流行的应用，通过源码解析将恶意代码植入到程序中再进行发布，这样的病毒普通用户无法辨别，因此下载应用最好在知名度较高的应用市场下载。

5.2 数据库操作

手机杀毒模块的核心在于病毒数据库的操作，根据病毒数据库中存储的 MD5 码与应用的 MD5 码进行对比，如果匹配则该应用是病毒。本节将针对病毒数据库的操作进行详细讲解。

5.2.1 数据库展示

随着智能手机的发展，手机病毒种类越来越多，病毒数据库中的数据也越来越庞大，为了方便，这里直接使用较完整的第三方病毒数据库 anvitirus.db。使用 SQLite Expert 打开病毒数据库的 datable 表，该表就用于存放病毒的相关信息，具体如图 5-3 所示。

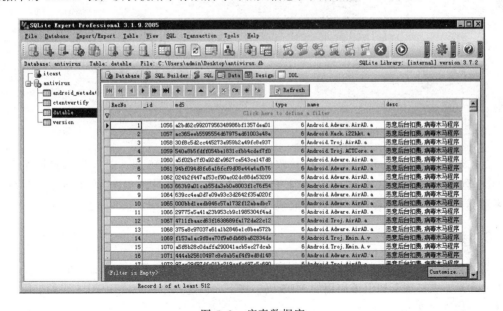

图 5-3　病毒数据库

如图 5-3 所示，datable 表中主要包含_id、md5、type、name、desc 等字段，这些字段分别用于表示自增主键、病毒 MD5 码、病毒类型、病毒名称、病毒描述信息。

5.2.2 数据库操作

将数据库复制到 assets 目录下，复制成功后，通过获取手机应用的 MD5 码与病毒数据库中的 MD5 码进行比对，如果当前应用的 MD5 码在病毒数据库中，则说明该应用就是病毒，这个过程实质上就是病毒查杀的过程，具体代码如【文件 5-1】所示。

【文件 5-1】AntiVirusDao.java

```java
1  public class AntiVirusDao{
2      /**
3       * 检查某个md5是否是病毒
4       * @param md5
5       * @return null 代表扫描安全
6       */
7      public static String checkVirus(String md5){
8          String desc=null;
9          //打开病毒数据库
10         SQLiteDatabase db=SQLiteDatabase.openDatabase(
11             "/data/data/cn.itcast.mobliesafe/files/antivirus.db", null,
12             SQLiteDatabase.OPEN_READONLY);
13         Cursor cursor=db.rawQuery("select desc from datable where md5=?",
14             new String[] {md5});
15         if(cursor.moveToNext()){
16             desc=cursor.getString(0);
17         }
18         cursor.close();
19         db.close();
20         return desc;
21     }
22 }
```

在上述代码中，checkVirus()方法接收的参数为应用的 MD5 码，打开病毒数据库，得到 MD5 列表，将参数中的 MD5 码与列表中的进行比对，如果该 MD5 码在列表中存在，则表示当前应用是病毒程序，然后获取病毒描述信息并返回。

5.2.3 获取 MD5 码

通过前面讲解可知，病毒数据库中保存的是手机病毒的 MD5 码，如果需要判断某个应用是否为病毒，则必须要知道该应用的 MD5 码，接下来创建一个获取应用 MD5 码的工具类，具体代码如【文件 5-2】所示。

【文件 5-2】 MD5Utils.java

```java
1  public class MD5Utils{
2      /**
3       * 获取文件的md5值,
4       * @param path 文件的路径
5       * @return null 文件不存在
6       */
7      public static String getFileMd5(String path){
8          try{
9              MessageDigest digest=MessageDigest.getInstance("md5");
10             File file=new File(path);
11             FileInputStream fis=new FileInputStream(file);
12             byte[] buffer=new byte[1024];
13             int len=-1;
14             while((len=fis.read(buffer))!=-1){
15                 digest.update(buffer,0,len);
16             }
17             byte[] result=digest.digest();
18             StringBuilder sb=new StringBuilder();
19             for(byte b:result){
20                 int number=b & 0xff;
21                 String hex=Integer.toHexString(number);
22                 if(hex.length()==1){
23                     sb.append("0"+hex);
24                 }else{
25                     sb.append(hex);
26                 }
27             }
28             return sb.toString();
29         } catch (Exception e){
30             e.printStackTrace();
31             return null;
32         }
33     }
34 }
```

在上述代码中，getFileMd5()方法接收一个应用的全路径名，最后返回该应用的 MD5 码。网络上许多获取文件 MD5 码的工具内部就是这样实现的。

至此，手机杀毒模块所用到的数据库和工具类已准备完成，包括数据库操作、获取应用程序的 MD5 码，接下来进行界面的开发。

> **多学一招：获取 MD5 码工具**
>
> 市面上有很多获取 MD5 码的工具，例如 MD5Count.exe 就是其中的一个，其图标是一个骷髅头。在使用这个工具时，直接双击该图标运行程序，然后将要计算的文件直接拖到该界面中即可生成 MD5 码，如图 5-4 所示。
>
>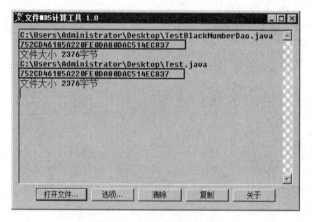
>
> 图 5-4　MD5 计算工具
>
> 需要注意的是，每个程序的 MD5 码都是唯一的。图 5-4 中的两个文件内容是相同的，只不过名字不同而已，因此获得到的 MD5 码是一样的。

5.3　病毒查杀

5.3.1　病毒查杀界面

病毒查杀界面主要由两部分组成，界面上方用于显示病毒图标以及上次扫描病毒的时间，下方用于显示全盘扫描病毒条目，点击该条目跳转到病毒查杀界面，病毒查杀的图形化界面如图 5-5 所示。

图 5-5 所示病毒查杀界面所对应的布局文件如【文件 5-3】所示。

【文件 5-3】activity_virusscan.xml

```
<?xml version="1.0" encoding="utf-8"?>
<LinearLayout xmlns:android="http://schemas.android.
    com/apk/res/android"
    android:layout_width="match_parent"
    android:layout_height="match_parent"
    android:orientation="vertical" >
    <include layout="@layout/titlebar"/>
    <RelativeLayout
        android:layout_width="match_parent"
        android:layout_height="180dp"
```

图 5-5　病毒查杀界面

```xml
        android:background="@color/blue">
        <ImageView
            android:layout_width="80dp"
            android:layout_height="80dp"
            android:layout_centerInParent="true"
            android:background="@drawable/virusscan_icon"/>
        <TextView
            android:id="@+id/tv_lastscantime"
            style="@style/textview16sp"
            android:layout_alignParentBottom="true"
            android:layout_centerHorizontal="true"
            android:layout_marginBottom="10dp"
            android:textColor="@color/white"/>
    </RelativeLayout>
    <RelativeLayout
        android:id="@+id/rl_allscanvirus"
        android:layout_width="match_parent"
        android:layout_height="70dp" >
        <ImageView
            android:id="@+id/imgv_quickscan"
            style="@style/wrapcontent"
            android:layout_alignParentLeft="true"
            android:layout_centerVertical="true"
            android:layout_marginLeft="10dp"
            android:src="@drawable/quickscan_icon"/>
        <TextView
            android:id="@+id/tv_allscan"
            style="@style/textview16sp"
            android:layout_marginLeft="15dp"
            android:layout_marginTop="15dp"
            android:layout_toRightOf="@+id/imgv_quickscan"
            android:text="全盘扫描"/>
        <TextView
            style="@style/textview14sp"
            android:layout_below="@+id/tv_allscan"
            android:layout_marginLeft="15dp"
            android:layout_marginTop="10dp"
            android:layout_toRightOf="@+id/imgv_quickscan"
            android:text="全面扫描手机，杀毒更彻底"/>
```

```xml
                <View
                    android:layout_width="13dp"
                    android:layout_height="20dp"
                    android:layout_alignParentRight="true"
                    android:layout_centerVertical="true"
                    android:layout_marginRight="15dp"
                    android:background="@drawable/rightarrow_blue"/>
                <View
                    android:layout_width="match_parent"
                    android:layout_height="1.0px"
                    android:layout_alignParentBottom="true"
                    android:background="@color/black30"/>
        </RelativeLayout>
</LinearLayout>
```

上述布局文件中,一个线性布局嵌套了两个相对布局,第一个相对布局中的 TextView 主要显示上次查杀的时间,如果之前没有查杀病毒则不显示;第二个相对布局显示病毒扫描的文字提示,该布局的主要作用是点击之后跳转页面,需要在代码中为该布局注册点击监听。

5.3.2　病毒查杀逻辑代码

病毒查杀界面逻辑较简单,主要功能是在初始化时将病毒数据库从 assets 目录下复制到工程目录中,并在页面获取焦点时通过 SharedPreferences 得到上次扫描的时间显示在界面上,具体代码如【文件 5-4】所示。

【文件 5-4】VirusScanActivity.Java

```java
1  public class VirusScanActivity extends Activity implements OnClickListener {
2      private TextView mLastTimeTV;
3      private SharedPreferences mSP;
4      @Override
5      protected void onCreate(Bundle savedInstanceState){
6          super.onCreate(savedInstanceState);
7          requestWindowFeature(Window.FEATURE_NO_TITLE);
8          setContentView(R.layout.activity_virusscan);
9          mSP=getSharedPreferences("config", MODE_PRIVATE);
10         copyDB("antivirus.db");
11         initView();
12     }
13     @Override
14     protected void onResume(){
15         String string=mSP.getString("lastVirusScan","您还没有查杀病毒!");
16         mLastTimeTV.setText(string);
```

```
17        super.onResume();
18    }
19    /**
20     * 复制病毒数据库
21     * @param string
22     */
23    private void copyDB(final String dbname){
24        new Thread(){
25            public void run(){
26                try{
27                    File file=new File(getFilesDir(),dbname);
28                    if(file.exists() && file.length()>0){
29                        Log.i("VirusScanActivity","数据库已存在! ");
30                        return;
31                    }
32                    InputStream is=getAssets().open(dbname);
33                    FileOutputStream fos=openFileOutput(dbname,MODE_PRIVATE);
34                    byte[] buffer=new byte[1024];
35                    int len=0;
36                    while((len=is.read(buffer))!=-1){
37                        fos.write(buffer,0,len);
38                    }
39                    is.close();
40                    fos.close();
41                } catch (Exception e){
42                    e.printStackTrace();
43                }
44            };
45        }.start();
46    }
47    /**
48     * 初始化控件
49     */
50    private void initView(){
51        findViewById(R.id.rl_titlebar).setBackgroundColor(
52            getResources().getColor(R.color.light_blue));
53        ImageView mLeftImgv=(ImageView) findViewById(R.id.imgv_leftbtn);
54        ((TextView) findViewById(R.id.tv_title)).setText("病毒查杀");
55        mLeftImgv.setOnClickListener(this);
56        mLeftImgv.setImageResource(R.drawable.back);
```

```
57      mLastTimeTV = (TextView) findViewById(R.id.tv_lastscantime);
58      findViewById(R.id.rl_allscanvirus).setOnClickListener(this);
59   }
60   @Override
61   public void onClick(View v){
62      switch(v.getId()){
63      case R.id.imgv_leftbtn:
64         finish();
65         break;
66      case R.id.rl_allscanvirus:
67         startActivity(new Intent(this,VirusScanSpeedActivity.class));
68         break;
69      }
70   }
71 }
```

代码说明：

- 第 10 行代码执行数据库的复制操作，将 assets 目录下的数据库复制到工程目录下。
- 第 14～18 行的 onResume()方法使用 SharedPreferences 对象的 getString()方法获取 key 为 lastVirusScan 的值，如果没有查杀过病毒，则将第二个参数的值显示在控件上，否则显示上次查杀的时间。
- 第 23～46 行的 copyDB()方法用于复制数据库，开启一个子线程通过 IO 流的形式将病毒数据库复制到当前目录下。
- 第 61～70 行的 onClick()方法为该界面按钮的点击事件，其中第 66～68 行代码为全盘扫描条目的点击事件，点击之后跳转至病毒查杀进度界面开始查杀病毒。

5.4 查杀进度

5.4.1 查杀进度界面

病毒查杀进度界面主要用于显示当前病毒查杀进度以及正在扫描的应用列表。该界面由三部分组成，上方按百分比显示查杀进度，中间 ListView 控件展示扫描的应用列表，下方是取消扫描按钮，查杀进度的图形化界面如图 5-6 所示。

图 5-6 所示查杀进度界面所对应的布局文件如【文件 5-5】所示。

【文件 5-5】activity_virusscanspeed.xml

```
<?xml version="1.0" encoding="utf-8"?>
<LinearLayout xmlns:android="http://schemas.android.
   com/apk/res/android"
   android:layout_width="match_parent"
```

图 5-6 病毒查杀进度界面

```xml
        android:layout_height="match_parent"
        android:orientation="vertical">
    <include layout="@layout/titlebar"/>
    <RelativeLayout
        android:layout_width="match_parent"
        android:layout_height="180dp"
        android:background="@color/blue">
        <TextView
            android:id="@+id/tv_scanprocess"
            style="@style/wrapcontent"
            android:layout_centerInParent="true"
            android:textColor="@color/white"
            android:textSize="38sp"
            android:textStyle="bold"/>
        <LinearLayout
            android:layout_width="match_parent"
            android:layout_height="wrap_content"
            android:layout_alignParentBottom="true"
            android:layout_marginBottom="10dp"
            android:layout_marginLeft="10dp"
            android:gravity="center_vertical"
            android:orientation="horizontal">
            <ImageView
                android:id="@+id/imgv_scanningicon"
                android:layout_width="30dp"
                android:layout_height="30dp"
                android:background="@drawable/scanning_icon"/>
            <TextView
                android:id="@+id/tv_scansapp"
                style="@style/textview16sp"
                android:layout_marginLeft="5dp"
                android:singleLine="true"
                android:textColor="@color/white"/>
        </LinearLayout>
    </RelativeLayout>
    <ListView
        android:id="@+id/lv_scanapps"
        android:layout_width="match_parent"
        android:layout_height="wrap_content"
```

```
        android:layout_weight="10"/>
    <LinearLayout
        android:layout_width="match_parent"
        android:layout_height="wrap_content"
        android:layout_margin="5dp"
        android:gravity="center" >
        <Button
            android:id="@+id/btn_canclescan"
            android:layout_width="312dp"
            android:layout_height="40dp"
            android:background="@drawable/cancle_scan_btn_selector"/>
    </LinearLayout>
</LinearLayout>
```

上述布局文件中，上方由一个相对布局包裹，中间包含了一个 TextView 和一个线性布局，线性布局中又包含一个 TextView 和一个 ImageView，其中第一个 TextView 显示正在查杀的进度，该进度由数字显示，达到 100%说明扫描完成，第二个 TextView 显示当前正在扫描的应用名称，根据应用的变化而变化。中间部分的 ListView 控件用于显示已扫描的应用列表。底部的相对布局中放置一个 Button 按钮用于取消扫描，该按钮使用了一个背景选择器，具体代码如【文件 5-6】所示。

【文件 5-6】res/drawable/cancle_scan_btn_selector.xml

```
<?xml version="1.0" encoding="utf-8"?>
<selector xmlns:android="http://schemas.android.com/apk/res/android">
    <item android:state_pressed="true" android:drawable="@drawable/cancle_scan_p"/>
    <item android:state_pressed="false" android:drawable="@drawable/cancle_scan_n"/>
</selector>
```

上述文件用于控制按钮按下与弹起时的图片，当按钮按下时显示 cancle_scan_p.png（蓝色）图片，当按钮弹起时显示 cancle_scan_n.png（灰色）图片。

5.4.2 查杀进度 Item 布局

由于病毒查杀界面中使用了 ListView 控件，因此需要定义一个 Item 布局用于展示应用程序信息，由于该布局与程序锁布局类似，因此可以直接使用程序锁布局，然后在代码中动态添加程序图标、名称、扫描图标即可，具体代码如【文件 5-7】所示。

【文件 5-7】item_list_applock.xml

```
<?xml version="1.0" encoding="utf-8"?>
<RelativeLayout xmlns:android="http://schemas.android.com/apk/res/android"
    android:layout_width="match_parent"
    android:layout_height="wrap_content"
    android:orientation="vertical">
```

```xml
<ImageView
    android:id="@+id/imgv_appicon"
    android:layout_width="50dp"
    android:layout_height="50dp"
    android:layout_alignParentLeft="true"
    android:layout_centerVertical="true"
    android:layout_margin="5dp"/>
<TextView
    android:id="@+id/tv_appname"
    style="@style/textview16sp"
    android:layout_centerVertical="true"
    android:layout_marginLeft="10dp"
    android:layout_toRightOf="@+id/imgv_appicon"
    android:textColor="@color/dark_gray"/>
<ImageView
    android:id="@+id/imgv_lock"
    android:layout_width="wrap_content"
    android:layout_height="wrap_content"
    android:layout_alignParentRight="true"
    android:layout_centerVertical="true"
    android:layout_marginRight="10dp"/>
</RelativeLayout>
```

上述布局中,使用了两个 ImageView 控件和一个 TextView 控件,其中第一个 ImageView 控件用于展示程序图标,第二个 ImageView 控件用于显示程序扫描完的图标,TextView 控件用于显示程序名称。

5.4.3 病毒的实体类

由于 ListView 列表显示的是手机中应用程序列表,其中包括应用名称、包名、应用图标、应用描述以及该应用是否是病毒。因此,首先需要根据这些信息定义一个病毒的实体类,具体代码如【文件 5-8】所示。

【文件 5-8】ScanAppInfo.java

```java
1  public class ScanAppInfo {
2      public String appName;
3      public boolean isVirus;
4      public String packagename;
5      public String description;
6      public Drawable appicon;
7  }
```

5.4.4 查杀进度逻辑

查杀进度界面主要功能是显示查杀进度的百分比以及正在扫描的程序,并且显示已扫描程序的列表,具体如【文件 5-9】所示。

【文件 5-9】VirusScanSpeedActivity.java

```
1  public class VirusScanSpeedActivity extends Activity implements
2  OnClickListener{
3      protected static final int SCAN_BENGIN=100;
4      protected static final int SCANNING=101;
5      protected static final int SCAN_FINISH=102;
6      private int total;
7      private int process;
8      private TextView mProcessTV;
9      private PackageManager pm;
10     private boolean flag;
11     private boolean isStop;
12     private TextView mScanAppTV;
13     private Button mCancleBtn;
14     private ImageView mScanningIcon;
15     private RotateAnimation rani;
16     private ListView mScanListView;
17     private ScanVirusAdapter adapter;
18     private List<ScanAppInfo> mScanAppInfos = new ArrayList<ScanAppInfo>();
19     private SharedPreferences mSP;
20     private Handler mHandler=new Handler(){
21         public void handleMessage(android.os.Message msg){
22             switch (msg.what){
23             case SCAN_BENGIN:
24                 mScanAppTV.setText("初始化杀毒引擎中...");
25                 break;
26             case SCANNING:
27                 ScanAppInfo info=(ScanAppInfo) msg.obj;
28                 mScanAppTV.setText("正在扫描: "+info.appName);
29                 int speed=msg.arg1;
30                 mProcessTV.setText((speed*100/total)+"%");
31                 mScanAppInfos.add(info);
32                 adapter.notifyDataSetChanged();
33                 mScanListView.setSelection(mScanAppInfos.size());
34                 break;
```

```
35          case SCAN_FINISH:
36              mScanAppTV.setText("扫描完成! ");
37              mScanningIcon.clearAnimation();
38              mCancleBtn.setBackgroundResource(R.drawable.scan_complete);
39              saveScanTime();
40              break;
41          }
42      }
43      private void saveScanTime(){
44          Editor edit=mSP.edit();
45          SimpleDateFormat sdf = new SimpleDateFormat("yyyy-MM-dd HH:mm:ss",
46          Locale.getDefault());
47          String currentTime=sdf.format(new Date());
48          currentTime="上次查杀: "+currentTime;
49          edit.putString("lastVirusScan",currentTime);
50          edit.commit();
51      };
52    };
53    @Override
54    protected void onCreate(Bundle savedInstanceState){
55        super.onCreate(savedInstanceState);
56        requestWindowFeature(Window.FEATURE_NO_TITLE);
57        setContentView(R.layout.activity_virusscanspeed);
58        pm=getPackageManager();
59        mSP=getSharedPreferences("config", MODE_PRIVATE);
60        initView();
61        scanVirus();
62    }
63    /**
64     * 扫描病毒
65     */
66    private void scanVirus() {
67        flag=true;
68        isStop=false;
69        process=0;
70        mScanAppInfos.clear();
71        new Thread(){
72            public void run() {
73                Message msg=Message.obtain();
```

```
74          msg.what=SCAN_BENGIN;
75          mHandler.sendMessage(msg);
76          List<PackageInfo> installedPackages=pm.getInstalledPackages(0);
77          total=installedPackages.size();
78          for(PackageInfo info:installedPackages){
79              if(!flag){
80                  isStop=true;
81                  return;
82              }
83              String apkpath=info.applicationInfo.sourceDir;
84              //检查获取这个文件的特征码
85              String md5info=MD5Utils.getFileMd5(apkpath);
86              String result=AntiVirusDao.checkVirus(md5info);
87              msg=Message.obtain();
88              msg.what=SCANNING;
89              ScanAppInfo scanInfo=new ScanAppInfo();
90              if(result==null){
91                  scanInfo.description="扫描安全";
92                  scanInfo.isVirus=false;
93              } else {
94                  scanInfo.description=result;
95                  scanInfo.isVirus=true;
96              }
97              process++;
98              scanInfo.packagename=info.packageName;
99              scanInfo.appName=info.applicationInfo.loadLabel(pm).
100             toString();
101             scanInfo.appicon=info.applicationInfo.loadIcon(pm);
102             msg.obj=scanInfo;
103             msg.arg1=process;
104             mHandler.sendMessage(msg);
105             try{
106                 Thread.sleep(300);
107             } catch(InterruptedException e){
108                 e.printStackTrace();
109             }
110         }
111         msg=Message.obtain();
112         msg.what=SCAN_FINISH;
```

```
113             mHandler.sendMessage(msg);
114         };
115     }.start();
116 }
117 private void initView(){
118     findViewById(R.id.rl_titlebar).setBackgroundColor(
119     getResources().getColor(R.color.light_blue));
120     ImageView mLeftImgv=(ImageView) findViewById(R.id.imgv_leftbtn);
121     ((TextView) findViewById(R.id.tv_title)).setText("病毒查杀进度");
122     mLeftImgv.setOnClickListener(this);
123     mLeftImgv.setImageResource(R.drawable.back);
124     mProcessTV=(TextView) findViewById(R.id.tv_scanprocess);
125     mScanAppTV=(TextView) findViewById(R.id.tv_scansapp);
126     mCancleBtn=(Button) findViewById(R.id.btn_canclescan);
127     mCancleBtn.setOnClickListener(this);
128     mScanListView=(ListView) findViewById(R.id.lv_scanapps);
129     adapter=new ScanVirusAdapter(mScanAppInfos,this);
130     mScanListView.setAdapter(adapter);
131     mScanningIcon=(ImageView) findViewById(R.id.imgv_scanningicon);
132     startAnim();
133 }
134 private void startAnim(){
135     if (rani==null){
136         rani=new RotateAnimation(0,360, Animation.RELATIVE_TO_SELF,
137         0.5f,Animation.RELATIVE_TO_SELF,0.5f);
138     }
139     rani.setRepeatCount(Animation.INFINITE);
140     rani.setDuration(2000);
141     mScanningIcon.startAnimation(rani);
142 }
143 @Override
144 public void onClick(View v){
145     switch(v.getId()){
146     case R.id.imgv_leftbtn:
147         finish();
148         break;
149     case R.id.btn_canclescan:
150         if(process==total & process>0) {
151             //扫描已完成
```

```
152            finish();
153        } else if(process>0 & process<total & isStop==false){
154            mScanningIcon.clearAnimation();
155            // 取消扫描
156            flag=false;
157            //更换背景图片
158            mCancleBtn.setBackgroundResource(R.drawable.restart_scan_btn);
159        } else if(isStop){
160            startAnim();
161            //重新扫描
162            scanVirus();
163            //更换背景图片
164            mCancleBtn.setBackgroundResource(
165                R.drawable.cancle_scan_btn_selector);
166        }
167        break;
168    }
169 }
170 @Override
171 protected void onDestroy(){
172    flag=false;
173    super.onDestroy();
174 }
175}
```

代码说明：

- 第 20~42 行代码定义了一个 Handler，该 Handler 接收三条从 scanVirus()方法中传输的数据，其中第 26~34 行代码接收 ScanAppInfo 信息，并将其中存储的应用信息显示在界面上，第 31 行代码在 mScanAppInfos 集合中不断添加数据，因此集合中的长度不断增加，在第 33 行代码使用 mScanListView.setSelection(mScanAppInfos.size())方法，使 ListView 一直选择最后一条数据，这样当不断扫描应用时，就会看到 ListView 自动向上滚动的效果。
- 第 54~62 行的 onCreate()方法是界面初始化时做的操作，分别是获取应用管理器、SharedPreferences 对象、初始化布局控件以及调用查杀病毒方法开始查杀。
- 第 66~116 行的 scanVirus()方法是扫描病毒的核心代码。该方法开启一个子线程，在子线程中处理扫描应用是否是病毒的操作，并将结果使用 Handler 发送进行处理。
- 第 73~77 行代码先向 Handler 发送一条 Message 表示开始准备杀毒；然后使用初始化操作中得到的应用程序管理器得到手机中安装的应用信息并存入集合 installedPackages 中，该集合包含了手机上安装的应用的基本信息，并定义一个变量 total 得到该集合的长度也就是集合中应用的个数。

- 第 78～82 行代码使用 for 循环遍历 installedPackages 集合，获取集合中应用的全路径名，然后调用 MD5Utils 类中的 getFileMd5(String md5)方法，在方法参数中传入应用的全路径名得到该应用的 MD5 码，然后将得到的 MD5 码作为参数调用 AntiVirusDao 类的 checkVirus()方法判断该 MD5 码是否存在病毒数据库中，如果该方法的返回值为空则说明该 MD5 所对应的应用程序不是病毒。
- 第 98～104 行代码得到应用信息存入 ScanAppInfo 中，然后发送给 Handler。
- 第 134～142 行的 startAnim()方法定义了一个旋转动画，用在布局的 ImageView 中，当开始扫描时，ImageView 进行顺时针的旋转。

5.4.5 数据适配器

病毒查杀界面中，使用了 ListView 控件，该控件中需要使用适配器进行填充，因此需要定义一个数据适配器类，具体代码如【文件 5-10】所示。

【文件 5-10】ScanVirusAdapter.java

```
1  public class ScanVirusAdapter extends BaseAdapter{
2      private List<ScanAppInfo> mScanAppInfos;
3      private Context context;
4      public ScanVirusAdapter(List<ScanAppInfo> scanAppInfo,Contextcontext){
5          super();
6          mScanAppInfos=scanAppInfo;
7          this.context=context;
8      }
9      @Override
10     public int getCount(){
11         return mScanAppInfos.size();
12     }
13     @Override
14     public Object getItem(int position){
15         return mScanAppInfos.get(position);
16     }
17     @Override
18     public long getItemId(int position){
19         return position;
20     }
21     @Override
22     public View getView(int position, View convertView, ViewGroup parent) {
23         ViewHolder holder;
24         if(convertView==null){
25             //由于程序锁的条目与病毒扫描内容基本一致，因此可以重用程序锁的 Item 布局
```

```java
26         convertView=View.inflate(context,R.layout.item_list_applock, null);
27         holder=new ViewHolder();
28         holder.mAppIconImgv=(ImageView) convertView.
29         findViewById(R.id.imgv_appicon);
30         holder.mAppNameTV=(TextView) convertView.
31         findViewById(R.id.tv_appname);
32         holder.mScanIconImgv=(ImageView) convertView.
33         findViewById(R.id.imgv_lock);
34         convertView.setTag(holder);
35     } else{
36         holder=(ViewHolder) convertView.getTag();
37     }
38     ScanAppInfo scanAppInfo=mScanAppInfos.get(position);
39     if(!scanAppInfo.isVirus){
40         holder.mScanIconImgv.setBackgroundResource(R.drawable.blue_right_icon);
41         holder.mAppNameTV.setTextColor(context.getResources().getColor(
42         R.color.black));
43         holder.mAppNameTV.setText(scanAppInfo.appName);
44     }else{
45         holder.mAppNameTV.setTextColor(context.getResources().getColor(
46         R.color.bright_red));
47         holder.mAppNameTV.setText(scanAppInfo.appName+"("+
48         scanAppInfo.description+")");
49     }
50     holder.mAppIconImgv.setImageDrawable(scanAppInfo.appicon);
51     return convertView;
52 }
53 static class ViewHolder {
54     ImageView mAppIconImgv;
55     TextView mAppNameTV;
56     ImageView mScanIconImgv;
57 }
58 }
```

在上述代码中，ScanVirusAdapter 类的构造方法接收一个 List<ScanAppInfo>参数，该参数保存了已经扫描过的应用程序，最后通过 getView()方法显示在 ListView 的条目中。

至此，手机杀毒模块的全部模块已经开发完成，编程者可以自行测试。

本 章 小 结

本章主要是针对手机杀毒模块进行讲解，首先讲解了病毒的特点与查杀方法，然后讲解病毒数据库的操作以及如何获取 MD5 码。最后讲解了手机杀毒模块的布局以及逻辑代码的开发。通过该模块的学习，可以让编程者掌握如何使用第三方数据库以及如何进行病毒查杀。

【面试精选】
1. 请问 StringBuffer 和 StringBuilder 有哪些区别？
2. 请问 Android 中有几类动画，它们的特点是什么？
扫描右方二维码，查看面试题答案！

第 6 章 缓存清理模块

学习目标
- 了解缓存清理模块功能；
- 掌握如何获取程序的缓存信息；
- 掌握如何清理程序的缓存。

随着手机的使用时间增长，手机中的缓存信息也就越多，这些缓存信息会导致手机运行速度变慢、顿卡，而且还会大大占用内存空间。因此，需要定时清理手机缓存，以保持手机状态良好。本章将针对缓存清理模块进行详细讲解。

6.1 模 块 概 述

6.1.1 功能介绍

缓存清理模块主要用于清理所有程序缓存，当扫描完所有程序后出现缓存时，可以点击"一键清理"按钮，此时会跳转到清理界面逐渐清理，清理完成后会显示成功清理了多少缓存。该模块界面效果如图 6-1 所示。

图 6-1 缓存清理界面

6.1.2 代码结构

缓存清理模块的核心代码就是获取所有程序的缓存信息，展示在 ListView 控件中，然后针对这些缓存进行清理，接下来通过一个图例来展示缓存清理模块的代码结构，如图 6-2 所示。

下面按照结构顺序依次介绍 chapter06 包中的文件，具体如下：

- CacheCleanAdapter.java：扫描缓存界面的 ListView 适配器。
- CacheInfo.java：用于存储缓存信息的实体类。
- CacheClearListActivity.java：扫描缓存程序界面的逻辑代码。
- CleanCacheActivity.java：缓存清理界面的逻辑代码。

图 6-2 chapter06 代码结构

6.2 扫 描 缓 存

手机中的大部分程序都有缓存信息，这些缓存信息是通过 AIDL 接口调用系统底层方法获取的。本节将针对扫描缓存功能进行详细讲解。

6.2.1 扫描缓存界面

扫描缓存界面与查杀病毒界面类似，都是由三部分组成：扫描缓存界面上方是一个扫把图标，不停地左右摆动表示在扫描缓存；中间是 ListView 控件，用于展示扫描完的程序；最下方是"一键清理"按钮。扫描缓存的图形化界面如图 6-3 所示。

图 6-3 扫描缓界面对应的布局文件如【文件 6-1】所示。

【文件 6-1】activity_cacheclearlist.xml

图 6-3 扫描缓存界面

```xml
<?xml version="1.0" encoding="utf-8"?>
<LinearLayout xmlns:android="http://schemas.android. com/apk/res/android"
    android:layout_width="match_parent"
    android:layout_height="match_parent"
    android:orientation="vertical" >
    <include layout="@layout/titlebar"/>
    <RelativeLayout
        android:layout_width="match_parent"
        android:layout_height="180dp"
        android:background="@color/light_rose_red" >
        <ImageView
            android:id="@+id/imgv_broom"
            android:layout_width="80dp"
            android:layout_height="80dp"
            android:layout_centerInParent="true"
```

```xml
            android:background="@anim/broom_animation" />
        <TextView
            android:id="@+id/tv_recommend_clean"
            style="@style/textview12sp"
            android:layout_width="150dp"
            android:layout_alignParentBottom="true"
            android:layout_alignParentLeft="true"
            android:layout_marginBottom="10dp"
            android:layout_marginLeft="10dp"
            android:maxLength="26"
            android:singleLine="true"
            android:textColor="@color/white"/>
        <TextView
            android:id="@+id/tv_can_clean"
            style="@style/textview12sp"
            android:layout_alignParentBottom="true"
            android:layout_alignParentRight="true"
            android:layout_marginBottom="10dp"
            android:layout_marginRight="10dp"
            android:maxLength="20"
            android:textColor="@color/white"/>
    </RelativeLayout>
    <ListView
        android:id="@+id/lv_scancache"
        android:layout_width="match_parent"
        android:layout_height="wrap_content"
        android:layout_weight="10"/>
    <LinearLayout
        android:layout_width="match_parent"
        android:layout_height="wrap_content"
        android:layout_margin="5dp"
        android:gravity="center">
        <Button
            android:id="@+id/btn_cleanall"
            android:layout_width="312dp"
            android:layout_height="40dp"
            android:background="@drawable/cleancache_btn_selector"
            android:enabled="false"/>
    </LinearLayout>
```

```
    </LinearLayout>
```

上述布局文件中，上方的相对布局包含一个 ImageView 和两个 TextView 控件，其中 ImageView 用于展示扫描图标（扫把），第一个 TextView 显示正在扫描的程序，第二个 TextView 显示总缓存大小。中间的 ListView 控件用于显示已扫描缓存的应用列表。下方的线性布局中放置一个 Button 按钮用于点击之后进行清理缓存操作。

上述布局中的 ImageView 控件使用了一个帧动画，类似 GIF 图片，通过一系列 Drawable 依次显示来模拟动画的效果，每张图片显示 0.2 秒进行切换，利用人眼视觉停留效果让人觉得扫把左右摆动清理缓存，具体代码如【文件 6-2】所示。

【文件 6-2】res/anim/broom_animation.xml

```xml
<?xml version="1.0" encoding="utf-8"?>
<animation-list xmlns:android="http://schemas.android.com/apk/res/android" >
    <item android:drawable="@drawable/broom_left" android:duration="200" />
    <item android:drawable="@drawable/broom_center" android:duration="200" />
    <item android:drawable="@drawable/broom_right" android:duration="200" />
</animation-list>
```

除了 ImageView 控件使用了背景选择器，"一键清理"按钮也使用了一个背景选择器，当按钮按下时显示 clean_all_p.png（粉色）图片，当按钮弹起时显示 clean_all_e.png（灰色）图片，具体代码如【文件 6-3】所示。

【文件 6-3】res/drawable/cleancache_btn_selector.xml

```xml
<?xml version="1.0" encoding="utf-8"?>
<selector xmlns:android="http://schemas.android.com/apk/res/android">
    <item android:state_pressed="true" android:drawable="@drawable/clean_all_p"/>
    <item android:state_enabled="false" android:drawable="@drawable/clean_all_e"/>
    <item android:drawable="@drawable/clean_all_n"/>
</selector>
```

6.2.2 缓存清理 Item 布局

扫描缓存界面下方使用了一个 ListView 控件，因此需要为其定义一个 Item 布局，该布局包含两个 ImageView 控件和两个 TextView 控件。其中第一个 ImageView 控件用于显示程序图标，第二个 ImageView 控件用于显示对号图标，第一个 TextView 用于显示程序名称，第二个 TextView 用于显示有多少缓存，具体代码如【文件 6-4】所示。

【文件 6-4】item_cacheclean_list.xml

```xml
<?xml version="1.0" encoding="utf-8"?>
<RelativeLayout xmlns:android="http://schemas.android.com/apk/res/android"
    android:layout_width="match_parent"
    android:layout_height="wrap_content"
    android:orientation="vertical">
    <ImageView
```

```xml
            android:id="@+id/imgv_appicon_cacheclean"
            android:layout_width="50dp"
            android:layout_height="50dp"
            android:layout_alignParentLeft="true"
            android:layout_centerVertical="true"
            android:layout_margin="5dp"/>
    <TextView
            android:id="@+id/tv_appname_cacheclean"
            style="@style/textview16sp"
            android:layout_marginLeft="10dp"
            android:layout_marginTop="10dp"
            android:layout_toRightOf="@+id/imgv_appicon_cacheclean"
            android:maxLength="24"/>
    <TextView
            android:id="@+id/tv_appsize_cacheclean"
            style="@style/textview12sp"
            android:layout_below="@+id/tv_appname_cacheclean"
            android:layout_marginLeft="10dp"
            android:layout_marginTop="5dp"
            android:layout_toRightOf="@+id/imgv_appicon_cacheclean"/>
    <ImageView
            android:layout_width="wrap_content"
            android:layout_height="wrap_content"
            android:layout_alignParentRight="true"
            android:layout_centerVertical="true"
            android:layout_marginRight="10dp"
            android:background="@drawable/cachescan_righticon"/>
</RelativeLayout>
```

6.2.3 缓存信息的实体类

由于 ListView 列表展示的是手机缓存信息，这些信息包括应用包名、缓存大小、应用图标及该应用名称。因此，首先需要根据这些信息定义一个缓存信息的实体类，具体代码如【文件 6-5】所示。

【文件 6-5】CacheInfo.java

```java
public class CacheInfo{
    public String packagename;
    public long cacheSize;
    public Drawable appIcon;
    public String appName;
```

```
}
```

6.2.4 扫描缓存逻辑

扫描缓存的主要逻辑是获取每个应用程序的缓存大小，然后将所有缓存累加在一起展示在界面上，并展示扫描后的应用列表，具体代码如【文件6-6】所示。

【文件6-6】CacheClearListActivity.java

```java
1  public class CacheClearListActivity extends Activity implements OnClickListener {
2      protected static final int SCANNING=100;
3      protected static final int FINISH=101;
4      private AnimationDrawable animation;
5      /** 建议清理 */
6      private TextView mRecomandTV;
7      /** 可清理 */
8      private TextView mCanCleanTV;
9      private long cacheMemory;
10     private List<CacheInfo> cacheInfos=new ArrayList<CacheInfo>();
11     private List<CacheInfo> mCacheInfos=new ArrayList<CacheInfo>();
12     private PackageManager pm;
13     private CacheCleanAdapter adapter;
14     private ListView mCacheLV;
15     private Button mCacheBtn;
16     private Handler handler=new Handler(){
17         public void handleMessage(Message msg){
18             switch(msg.what){
19             case SCANNING:
20                 PackageInfo info=(PackageInfo) msg.obj;
21                 mRecomandTV.setText("正在扫描: "+info.packageName);
22                 mCanCleanTV.setText("已扫描缓存 : "+Formatter.formatFileSize(
23                     CacheClearListActivity.this,cacheMemory));
24                 //在主线程添加变化后集合
25                 mCacheInfos.clear();
26                 mCacheInfos.addAll(cacheInfos);
27                 //ListView 刷新
28                 adapter.notifyDataSetChanged();
29                 mCacheLV.setSelection(mCacheInfos.size());
30                 break;
31             case FINISH:
32                 //扫描完毕,动画停止
33                 animation.stop();
```

```java
34          if(cacheMemory>0){
35              mCacheBtn.setEnabled(true);
36          }else{
37              mCacheBtn.setEnabled(false);
38              Toast.makeText(CacheClearListActivity.this,"您的手机洁净如新",0).
39              show();
40          }
41          break;
42      }
43    };
44  };
45  private Thread thread;
46  @Override
47  protected void onCreate(Bundle savedInstanceState){
48      super.onCreate(savedInstanceState);
49      requestWindowFeature(Window.FEATURE_NO_TITLE);
50      setContentView(R.layout.activity_cacheclearlist);
51      pm=getPackageManager();
52      initView();
53  }
54  /**
55   * 初始化控件
56   */
57  private void initView(){
58      findViewById(R.id.rl_titlebar).setBackgroundColor(
59          getResources().getColor(R.color.rose_red));
60      ImageView mLeftImgv=(ImageView) findViewById(R.id.imgv_leftbtn);
61      mLeftImgv.setOnClickListener(this);
62      mLeftImgv.setImageResource(R.drawable.back);
63      ((TextView) findViewById(R.id.tv_title)).setText("缓存扫描");
64      mRecomandTV=(TextView) findViewById(R.id.tv_recommend_clean);
65      mCanCleanTV=(TextView) findViewById(R.id.tv_can_clean);
66      mCacheLV=(ListView) findViewById(R.id.lv_scancache);
67      mCacheBtn=(Button) findViewById(R.id.btn_cleanall);
68      mCacheBtn.setOnClickListener(this);
69      animation=(AnimationDrawable) findViewById(R.id.imgv_broom).
70          getBackground();
71      animation.setOneShot(false);
72      animation.start();
```

```
73      adapter=new CacheCleanAdapter(this, mCacheInfos);
74      mCacheLV.setAdapter(adapter);
75      fillData();
76    }
77    /**
78     * 填充数据
79     */
80    private void fillData(){
81       thread=new Thread(){
82          public void run(){
83             //遍历手机中所有的应用程序
84             cacheInfos.clear();
85             List<PackageInfo> infos=pm.getInstalledPackages(0);
86             for(PackageInfo info:infos){
87                getCacheSize(info);
88                try{
89                   Thread.sleep(500);
90                }catch(InterruptedException e){
91                   e.printStackTrace();
92                }
93                Message msg=Message.obtain();
94                msg.obj=info;
95                msg.what=SCANNING;
96                handler.sendMessage(msg);
97             }
98             Message msg=Message.obtain();
99             msg.what=FINISH;
100            handler.sendMessage(msg);
101        };
102     };
103     thread.start();
104   }
105   /**
106    * 获取某个包名对应的应用程序的缓存大小
107    * @param info 应用程序的包信息
108    */
109   public void getCacheSize(PackageInfo info){
110      try{
111         Method method=PackageManager.class.getDeclaredMethod(
```

```
112              "getPackageSizeInfo", String.class, IPackageStatsObserver.class);
113          method.invoke(pm,info.packageName,new MyPackObserver(info));
114      }catch(Exception e){
115          e.printStackTrace();
116      }
117  }
118  @Override
119  protected void onDestroy(){
120      super.onDestroy();
121      animation.stop();
122      if(thread!=null){
123          thread.interrupt();
124      }
125  }
126  @Override
127  public void onClick(View v){
128      switch (v.getId()){
129      case R.id.imgv_leftbtn:
130          finish();
131          break;
132      case R.id.btn_cleanall:
133          if(cacheMemory>0){
134              //跳转至清理缓存的页面的Activity
135              Intent intent=new Intent(this,CleanCacheActivity.class);
136              //将要清理的垃圾大小传递至另一个页面
137              intent.putExtra("cacheMemory",cacheMemory);
138              startActivity(intent);
139              finish();
140          }
141          break;
142      }
143  }
144  private class MyPackObserver extends android.content.pm.
145  IPackageStatsObserver.Stub{
146      private PackageInfo info;
147      public MyPackObserver(PackageInfo info){
148          this.info=info;
149      }
150      @Override
```

```
151    public void onGetStatsCompleted(PackageStats pStats, boolean succeeded)
152      throws RemoteException{
153        long cachesize=pStats.cacheSize;
154        if(cachesize>=0){
155           CacheInfo cacheInfo=new CacheInfo();
156           cacheInfo.cacheSize=cachesize;
157           cacheInfo.packagename=info.packageName;
158           cacheInfo.appName=info.applicationInfo.loadLabel(pm).toString();
159           cacheInfo.appIcon=info.applicationInfo.loadIcon(pm);
160           cacheInfos.add(cacheInfo);
161           cacheMemory+=cachesize;
162        }
163      }
164    }
165 }
```

代码说明：

- 第 16~44 行代码定义了一个 Handler，用于接收消息判断程序是正在扫描还是已经扫描完成，当正在扫描时需要获取当前扫描的是哪个程序，以及展示已扫描完的程序列表。如果已经扫描完毕，则停止动画，判断缓存大小，如果缓存大于 0，则设置"一键清理"按钮可用，否则设置该按钮不可用，并弹出 Toast 显示手机洁净如新。
- 第 80~104 行的 fillData()方法用于填充数据，单开启一个线程，在该线程中遍历所有已安装的程序，并通过 Handler 发送消息，当正在遍历程序时，发送消息 SCANNING（正在扫描），遍历完成后发送消息 FINISH（扫描完成）。
- 第 109~117 行的 getCacheSize()方法用于获取每个程序的缓存大小，由于 getPackageSizeInfo (String packageName,IPackageStatsObserver observer)方法是隐藏的，因此需要通过反射机制获取该方法，该方法参数中的 packageName 表示包名，observer 是一个远程服务的 AIDL 接口。
- 第 144~164 行的 MyPackObserver 类是一个 AIDL 的实现类，它继承自 android.content.pm. IPackageStatsObserver.Stub，需要实现该类中的 onGetStatsCompleted()方法，然后通过 pStats.cacheSize()属性就可以获取到缓存信息，最后将扫描过的程序添加到 cacheInfos 集合中，并累计所有的程序缓存数据。

需要注意的是，获取缓存时需要使用 IPackageStatsObserver 接口，因此需要创建一个 android. content.pm 包，将 IPackageStatsObserver.aidl 复制到工程中。由于该接口还依赖于 PackageStats.aidl 接口，因此也需要将该文件复制到 android.content.pm 包中。

在获取程序缓存时，需要在 AndroidManifest.xml 文件中配置相关权限，具体代码如下：

```
<uses-permission android:name="android.permission.GET_PACKAGE_SIZE" />
```

6.2.5 数据适配器

由于扫描缓存界面使用了 ListView 控件，该控件中需要使用适配器进行填充，因此需要定义一个数据适配器类,具体代码如【文件6-7】所示。

【文件6-7】 CacheCleanAdapter.java

```java
public class CacheCleanAdapter extends BaseAdapter{
    private Context context;
    private List<CacheInfo> cacheInfos;
    public CacheCleanAdapter(Context context,List<CacheInfo> cacheInfos){
        super();
        this.context=context;
        this.cacheInfos=cacheInfos;
    }
    @Override
    public int getCount(){
        return cacheInfos.size();
    }
    @Override
    public Object getItem(int position){
        return cacheInfos.get(position);
    }
    @Override
    public long getItemId(int position){
        return position;
    }
    @Override
    public View getView(int position,View convertView,ViewGroup parent){
        ViewHolder holder=null;
        if (convertView==null){
            holder=new ViewHolder();
            convertView=View.inflate(context,R.layout.item_cacheclean_list,null);
            holder.mAppIconImgv=(ImageView) convertView.
                    findViewById(R.id.imgv_appicon_cacheclean);
            holder.mAppNameTV=(TextView) convertView.
                    findViewById(R.id.tv_appname_cacheclean);
            holder.mCacheSizeTV=(TextView) convertView.
                    findViewById(R.id.tv_appsize_cacheclean);
            convertView.setTag(holder);
        }else{
            holder=(ViewHolder) convertView.getTag();
        }
        CacheInfo cacheInfo=cacheInfos.get(position);
```

```
38        holder.mAppIconImgv.setImageDrawable(cacheInfo.appIcon);
39        holder.mAppNameTV.setText(cacheInfo.appName);
40        holder.mCacheSizeTV.setText(Formatter.formatFileSize(context,
41        cacheInfo.cacheSize));
42        return convertView;
43    }
44    static class ViewHolder{
45        ImageView mAppIconImgv;
46        TextView mAppNameTV;
47        TextView mCacheSizeTV;
48    }
49}
```

6.3 缓存清理

缓存清理功能主要是通过反射隐藏的方法 freeStorcegeAndNotify()对缓存进行清理，并利用了 Android 系统的一个漏洞。本节将针对缓存清理功能进行详细讲解。

6.3.1 缓存清理界面

缓存清理界面主要由两个帧布局组成，第一个帧布局用于展示正在清理缓存界面，第二个帧布局用于展示缓存清理完成界面，通过控制帧布局的显示与隐藏就可以展示当前在进行哪项操作。默认情况下清理完成的帧布局是隐藏的，这里为了让编程者看到更直观的效果，将其分别进行展示。缓存清理的图形化界面如图 6-4 所示。

图 6-4 缓存清理界面

图 6-4 所示缓存清理界面对应的布局文件如【文件 6-8】所示。

【文件 6-8】activity_cleancache.xml

```
<?xml version="1.0" encoding="utf-8"?>
<LinearLayout xmlns:android="http://schemas.android.com/apk/res/android"
    android:layout_width="match_parent"
```

```xml
    android:layout_height="match_parent"
    android:orientation="vertical">
    <include layout="@layout/titlebar"/>
    <RelativeLayout
        android:layout_width="match_parent"
        android:layout_height="match_parent"
        android:background="@color/light_rose_red">
        <FrameLayout
            android:id="@+id/fl_cleancache"
            android:layout_width="match_parent"
            android:layout_height="match_parent">
            <LinearLayout
                android:layout_width="match_parent"
                android:layout_height="match_parent"
                android:gravity="center"
                android:orientation="vertical">
                <ImageView
                    android:id="@+id/imgv_trashbin_cacheclean"
                    android:layout_width="wrap_content"
                    android:layout_height="wrap_content"
                    android:background="@anim/cacheclean_trash_bin_animation"/>
                <TextView
                    style="@style/wrapcontent"
                    android:layout_marginTop="15dp"
                    android:text="已清理垃圾文件: "
                    android:textColor="@color/white"
                    android:textSize="20sp"/>
                <LinearLayout
                    android:layout_width="wrap_content"
                    android:layout_height="wrap_content"
                    android:layout_marginTop="15dp"
                    android:orientation="horizontal">
                    <TextView
                        android:id="@+id/tv_cleancache_memory"
                        style="@style/wrapcontent"
                        android:textColor="@color/white"
                        android:textScaleX="1.2"
                        android:textSize="48sp"
                        android:textStyle="bold" />
```

```xml
            <TextView
                android:id="@+id/tv_cleancache_memoryunit"
                style="@style/wrapcontent"
                android:layout_marginLeft="5dp"
                android:textColor="@color/white"
                android:textSize="22sp" />
        </LinearLayout>
    </LinearLayout>
</FrameLayout>
<FrameLayout
    android:id="@+id/fl_finishclean"
    android:layout_width="match_parent"
    android:layout_height="match_parent"
    android:visibility="gone">
    <LinearLayout
        android:layout_width="match_parent"
        android:layout_height="match_parent"
        android:orientation="vertical" >
        <LinearLayout
            android:layout_width="match_parent"
            android:layout_height="match_parent"
            android:layout_weight="10"
            android:gravity="center"
            android:orientation="vertical" >
            <ImageView
                android:layout_width="wrap_content"
                android:layout_height="wrap_content"
                android:background="@drawable/cleancache_finish"/>
            <TextView
                android:id="@+id/tv_cleanmemorysize"
                style="@style/wrapcontent"
                android:layout_marginTop="30dp"
                android:textColor="@color/white"
                android:textSize="20sp"/>
        </LinearLayout>
        <LinearLayout
            android:layout_width="match_parent"
            android:layout_height="wrap_content"
            android:gravity="center">
```

```xml
            <Button
                android:id="@+id/btn_finish_cleancache"
                android:layout_width="wrap_content"
                android:layout_height="wrap_content"
                android:layout_gravity="center"
                android:layout_margin="10dp"
                android:background="@drawable/finish_cleancache_selector"/>
        </LinearLayout>
        </LinearLayout>
    </FrameLayout>
</RelativeLayout>
</LinearLayout>
```

上述布局文件中，第一个帧布局中的 ImageView 控件使用了一个动画选择器，该选择器是一个帧动画，用于控制垃圾桶图标 400 毫秒切换一次（一个图标中的垃圾桶是关着的，另一个是打开的，两张图片快速切换，给人以垃圾桶自动开关回收垃圾的感觉），具体代码如【文件 6-9】所示。

【文件 6-9】 res/anim/cacheclean_trash_bin_animation.xml

```xml
<?xml version="1.0" encoding="utf-8"?>
<animation-list xmlns:android="http://schemas.android.com/apk/res/android" >
    <item android:drawable="@drawable/cacheclean_trashbin_close_icon"
        android:duration="400"/>
    <item android:drawable="@drawable/cacheclean_trashbin_open_icon"
        android:duration="400"/>
</animation-list>
```

第二个帧布局中的"完成"按钮同样使用了一个背景选择器，当按钮按下时显示 finish_cleancache_p.png 图片（粉色），当按钮弹起时显示 finish_cleancache_n.png 图片（灰色），具体代码如【文件 6-10】所示。

【文件 6-10】 res/drawable/finish_cleancache_selector.xml

```xml
<?xml version="1.0" encoding="utf-8"?>
<selector xmlns:android="http://schemas.android.com/apk/res/android" >
    <item android:state_pressed="true"
            android:drawable="@drawable/finish_cleancache_p"/>
    <item android:state_pressed="false"
            android:drawable="@drawable/finish_cleancache_n"/>
</selector>
```

6.3.2 缓存清理逻辑

在清理缓存时，需要在界面上不停地更新数据，显示清理了多少缓存。当缓存清理完成后会在代码中动态切换布局，显示缓存清理完成界面。具体代码如【文件 6-11】所示。

【文件 6-11】CleanCacheActivity.java

```java
1  public class CleanCacheActivity extends Activity implements OnClickListener{
2      protected static final int CLEANNING=100;
3      private AnimationDrawable animation;
4      private long cacheMemory;
5      private TextView mMemoryTV;
6      private TextView mMemoryUnitTV;
7      private PackageManager pm;
8      private FrameLayout mCleanCacheFL;
9      private FrameLayout mFinishCleanFL;
10     private TextView mSizeTV;
11     private Handler mHandler=new Handler(){
12         public void handleMessage(android.os.Message msg){
13             switch (msg.what){
14             case CLEANNING:
15                 long memory=(Long) msg.obj;
16                 formatMemory(memory);
17                 if(memory==cacheMemory){
18                     animation.stop();
19                     mCleanCacheFL.setVisibility(View.GONE);
20                     mFinishCleanFL.setVisibility(View.VISIBLE);
21                     mSizeTV.setText("成功清理: "+Formatter.formatFileSize(
22                         CleanCacheActivity.this,cacheMemory));
23                 }
24                 break;
25             }
26         };
27     };
28     @Override
29     protected void onCreate(Bundle savedInstanceState){
30         super.onCreate(savedInstanceState);
31         requestWindowFeature(Window.FEATURE_NO_TITLE);
32         setContentView(R.layout.activity_cleancache);
33         initView();
34         pm=getPackageManager();
35         Intent intent=getIntent();
36         cacheMemory=intent.getLongExtra("cacheMemory",0);
37         initData();
```

```java
38    }
39    private void initView(){
40        findViewById(R.id.rl_titlebar).setBackgroundColor(
41            getResources().getColor(R.color.rose_red));
42        ((TextView) findViewById(R.id.tv_title)).setText("缓存清理");
43        ImageView mLeftImgv=(ImageView) findViewById(R.id.imgv_leftbtn);
44        mLeftImgv.setOnClickListener(this);
45        mLeftImgv.setImageResource(R.drawable.back);
46        animation=(AnimationDrawable) findViewById(
47            R.id.imgv_trashbin_cacheclean).getBackground();
48        animation.setOneShot(false);
49        animation.start();
50        mMemoryTV=(TextView) findViewById(R.id.tv_cleancache_memory);
51        mMemoryUnitTV=(TextView) findViewById(R.id.tv_cleancache_memoryunit);
52        mCleanCacheFL=(FrameLayout) findViewById(R.id.fl_cleancache);
53        mFinishCleanFL=(FrameLayout) findViewById(R.id.fl_finishclean);
54        mSizeTV=(TextView) findViewById(R.id.tv_cleanmemorysize);
55        findViewById(R.id.btn_finish_cleancache).setOnClickListener(this);
56    }
57    private void initData(){
58        cleanAll();
59        new Thread(){
60            public void run(){
61                long memory=0;
62                while(memory<cacheMemory){
63                    try{
64                        Thread.sleep(300);
65                    } catch(InterruptedException e){
66                        e.printStackTrace();
67                    }
68                    Random rand=new Random();
69                    int i=rand.nextInt();
70                    i=rand.nextInt(1024);
71                    memory+=1024*i;
72                    if(memory>cacheMemory){
73                        memory=cacheMemory;
74                    }
75                    Message message=Message.obtain();
76                    message.what=CLEANNING;
```

```
77              message.obj=memory;
78              mHandler.sendMessageDelayed(message,200);
79          }
80      };
81    }.start();
82  }
83  /**
84   * 初始化数据
85   */
86  private void formatMemory(long memory){
87      String cacheMemoryStr = Formatter.formatFileSize(this,memory);
88      String memoryStr;
89      String memoryUnit;
90      //根据大小判定单位
91      if(memory>900){
92          //大于900则单位是两位
93          memoryStr = cacheMemoryStr.substring(0, cacheMemoryStr.length()-2);
94          memoryUnit=cacheMemoryStr.substring(cacheMemoryStr.length()-2,
95          cacheMemoryStr.length());
96      }else{
97          //单位是一位
98          memoryStr=cacheMemoryStr.substring(0, cacheMemoryStr.length()-1);
99          memoryUnit=acheMemoryStr.substring(cacheMemoryStr.length()-1,
100         cacheMemoryStr.length());
101     }
102     mMemoryTV.setText(memoryStr);
103     mMemoryUnitTV.setText(memoryUnit);
104 }
105 @Override
106 public void onClick(View v){
107     switch (v.getId()){
108     case R.id.imgv_leftbtn:
109         finish();
110         break;
111     case R.id.btn_finish_cleancache:
112         finish();
113         break;
114     }
115 }
```

```
116    class ClearCacheObserver extends android.content.pm.
117    IPackageDataObserver.Stub{
118        public void onRemoveCompleted(final String packageName,
119        final boolean succeeded){
120        }
121    }
122    private void cleanAll(){
123        //清除全部缓存利用Android系统的一个漏洞: freeStorageAndNotify
124        Method[] methods=PackageManager.class.getMethods();
125        for(Method method:methods){
126            if("freeStorageAndNotify".equals(method.getName())){
127                try{
128                    method.invoke(pm, Long.MAX_VALUE,new ClearCacheObserver());
129                } catch(Exception e){
130                    e.printStackTrace();
131                }
132                return;
133            }
134        }
135        Toast.makeText(this, "清理完毕", 0).show();
136    }
137 }
```

代码说明:

- 第11~27行代码定义了一个Handler，用于处理从initData()方法中传递的数据。首先通过formatMemory()方法格式化单位，将获得的memory的值显示在界面上，显示正在清理缓存的过程。然后判断memory的值是否和缓存的值相等，如果相等说明已经清理完毕，此时停止缓存清理的动画，然后动态切换布局让清理完成的帧布局显示，并隐藏清理缓存的布局。

- 第35~37行代码通过Intent获得从上个界面传输过来的缓存总大小。

- 第39~56行 initView()方法为初始化布局，在onCreate()方法中调用，在第46~49行代码定义了一个动画，该动画为两个图片的动态切换，分别是一个打开的垃圾桶和盖上盖子的垃圾桶，通过这两张图片的不断切换展示出一种正在清理垃圾的效果。

- 第57~82行 initData()方法，在onCreate()方法中调用，首先执行了cleanAll()方法清除缓存，cleanAll()方法下面详细讲解；由于缓存清理速度很快，用户看不到清理的过程，因此开启一个子线程，首先定义一个long类型的变量memory初始化为0，判断如果该值小于缓存的值，就是用Random类随机生成一个范围在1024内的整数，将随机生成的数和1024相乘并将该值赋值给memory,然后判断memory的值是否大于缓存的值，如果大于则将缓存的值赋给memory,最后通过Message将memory发送给Handler对象处理。

- 第 86～104 行的 formatMemory()方法用于格式化单位,其中第 87 行代码 Formatter.formatFileSize(this,memory)将传入的 long 类型值转换为缓存文件的大小和单位。查看系统源码得知,Formatter.formatFileSize(this,memory)方法计算数值的单位是以 900 为界限的,如果数值小于 900,该数值的单位为 B,如果大于 900,可能为 KB、MB 或者 GB。因此判断接收的数值范围是否大于或者小于等于 900,以此来分割数字与单位。第 91～95 行代码,如果数值大于 900 代表该数的单位为 KB 以上,因此截取后两位字母,分别获取数字和单位;第 96～101 行代码数值小于或者等于 900,说明该数的单位为 B,因此截取后一位字母。第 102～103 行代码将截取的数字和单位分别显示在不同的文本控件中。

- 第 116～121 行的 ClearCacheObserver 类是远程服务的代理类,实现了 IpackageDataObserver 接口中的 onRemoveCompleted()方法。

- 第 122～136 行 cleanAll()方法用于清除所有程序缓存,通过反射的形式获取到 freeStorageAndNotify(long freeStorageSize, IPackageDataObserver observer)方法,该方法接收两个参数,第一个参数表示要释放的缓存大小,第二个参数是远程服务接口。其中 freeStorageSize 参数比较特殊,假设要释放 2 MB 缓存,手机有 5 MB 的内存空间,那么就直接清除 2 MB 的缓存;假设设置要释放 5 MB 的缓存,但手机实际只有 2 MB 的内存空间,那么就只能释放 2 MB 的缓存。这个 API 可以利用系统漏洞,将这个释放缓存的值设置得非常大,例如要释放 100 GB 的内存空间,手机只有 2 GB 内存,而缓存文件占 1 GB,系统发现即使把缓存文件都删除也无法释放那么多的内存,因此就把缓存文件都清理了。目前市面上的安全软件也都是这样做的。Google 检测该 API 时没有检测清除缓存权限,因此一般程序可以使用。

需要注意的是,在清除缓存时使用了 IPackageDataObserver.aidl 接口,因此需要将其复制到 android.content.pm 包中。另外,还要在 AndroidManifest.xml 文件中配置清除缓存相关权限,具体代码如下:

```
<uses-permission android:name="android.permission.CLEAR_APP_CACHE"/>
```

本 章 小 结

本章主要是针对缓存清理模块进行讲解,首先讲解了如何对程序进行扫描,获取到程序中缓存信息,然后讲解了如何清理缓存。该模块的界面效果与手机杀毒模块非常类似,编程者可以根据手机杀毒模块来设计该模块界面。

【面试精选】
1. 请问 AIDL 的工作原理以及实现步骤?
2. 请问接口和抽象类有哪些区别?
扫描右方二维码,查看面试题答案!

第 7 章

➡ 进程管理模块

学习目标

- 了解进程管理模块功能；
- 掌握如何获取手机中的进程；
- 掌握如何结束正在运行的进程。

大家都知道，大部分 Android 手机用的时间越长就会越慢甚至卡顿，哪怕是 2 GB 的运行内存，时间一久手机也会变得奇慢无比。通常情况下，造成这种情况的原因有两种，一种是缓存垃圾过多，另一种是后台开启的进程过多。第 6 章已经针对缓存清理进行了讲解，本章将针对进程管理进行详细讲解，以解决手机运行速度过慢问题。

7.1 模 块 概 述

7.1.1 功能介绍

进程管理模块主要用于查看当前开启多少进程服务，其中包含用户进程个数和系统进程个数，我们可以选择清理某个进程，也可以选择清理所有进程，同时还可以设置是否显示系统进程以及锁屏时是否清理进程。该模块界面效果如图 7-1 所示。

图 7-1 进程管理界面

7.1.2 代码结构

进程管理模块代码比较简单，主要包含进程管理界面及进程设置界面的逻辑代码，接下来通过一个图例来展示进程管理模块的代码结构，如图 7-2 所示。

图 7-2 chapter07 代码结构

下面按照结构顺序依次介绍 chapter07 包中的文件，具体如下：
- ProcessManagerAdapter.java：进程管理的数据适配器；
- TaskInfo.java：进程信息的实体类；
- AutoKillProcessService.java：锁屏清理进程的服务类；
- SystemInfoUtils.java：获取手机内存信息的工具类；
- TaskInfoParser.java：进程信息的解析器；
- ProcessManagerActivity.java：进程管理界面的逻辑代码；
- ProcessManagerSettingActivity.java：进程设置界面的逻辑代码。

7.2 进程管理

7.2.1 进程管理界面

进程管理模块界面主要由三个线性布局组成，第一个线性布局用于显示正在运行的进程个数、可用内存和总内存大小，第二个线性布局用于展示所有运行中的进程列表，第三个线性布局用于展示三个按钮，分别为清理、全选、反选，具体如图 7-3 所示。

图 7-3 所示进程管理界面对应的布局如【文件 7-1】所示。

【文件 7-1】activity_processmanager.xml

```
<?xml version="1.0" encoding="utf-8"?>
<LinearLayout xmlns:android="http://schemas.android.
    com/apk/res/android"
    android:layout_width="match_parent"
```

图 7-3 进程管理界面

```xml
        android:layout_height="match_parent"
        android:orientation="vertical" >
    <include layout="@layout/titlebar"/>
    <RelativeLayout
        android:layout_width="match_parent"
        android:layout_height="wrap_content"
        android:layout_margin="5dp" >
        <TextView
            android:id="@+id/tv_runningprocess_num"
            style="@style/textview12sp"
            android:layout_alignParentLeft="true"/>
        <TextView
            android:id="@+id/tv_memory_processmanager"
            style="@style/textview12sp"
            android:layout_alignParentRight="true"/>
    </RelativeLayout>
    <RelativeLayout
        android:layout_width="match_parent"
        android:layout_height="match_parent"
        android:layout_weight="10">
        <ListView
            android:id="@+id/lv_runningapps"
            android:layout_width="match_parent"
            android:layout_height="wrap_content"/>
        <TextView
            android:id="@+id/tv_user_runningprocess"
            android:layout_width="match_parent"
            android:layout_height="wrap_content"
            android:background="@color/graye5"
            android:padding="5dp"/>
    </RelativeLayout>
    <RelativeLayout
        android:layout_width="match_parent"
        android:layout_height="wrap_content"
        android:gravity="center_vertical"
        android:layout_marginBottom="10dp"
        android:layout_marginTop="10dp">
        <Button
            android:id="@+id/btn_cleanprocess"
```

```xml
            android:layout_width="90dp"
            android:layout_height="40dp"
            android:layout_alignParentLeft="true"
            android:layout_marginLeft="10dp"
            android:background="@drawable/cleanprocess_btn_selector"/>
        <Button
            android:id="@+id/btn_selectall"
            android:layout_width="90dp"
            android:layout_height="40dp"
            android:layout_centerHorizontal="true"
            android:background="@drawable/select_all_btn_selector"/>
        <Button
            android:id="@+id/btn_select_inverse"
            android:layout_width="90dp"
            android:layout_height="40dp"
            android:layout_alignParentRight="true"
            android:layout_marginRight="10dp"
            android:background="@drawable/inverse_btn_selector"/>
    </RelativeLayout>
</LinearLayout>
```

上述布局文件中，位于界面下方的清理、全选、反选三个按钮都使用了背景选择器，均用于控制按钮的按下与弹起效果，当按钮按下时显示绿色的背景图片，当按钮弹起时显示灰色的背景图片，接下来分别列举这三个背景选择器代码，具体如下所示：

（1）"清理"按钮的背景选择器，具体代码如【文件7-2】所示。

【文件7-2】res/drawable/cleanprocess_btn_selector.xml

```xml
<?xml version="1.0" encoding="utf-8"?>
<selector xmlns:android="http://schemas.android.com/apk/res/android">
    <item android:state_pressed="true"
                android:drawable="@drawable/clean_process_btn_p"/>
    <item android:state_pressed="false"
                android:drawable="@drawable/clean_process_btn_n"/>
</selector>
```

（2）"全选"按钮的背景选择器，具体代码如【文件7-3】所示。

【文件7-3】res/drawable/select_all_btn_selector.xml

```xml
<?xml version="1.0" encoding="utf-8"?>
<selector xmlns:android="http://schemas.android.com/apk/res/android">
    <item android:state_pressed="true" android:drawable="@drawable/ select_all_btn_p"/>
```

```xml
    <item android:state_pressed="false" android:drawable="@drawable/select_
    all_btn_n"/>
</selector>
```

(3)"反选"按钮的背景选择器,具体代码如【文件 7-4】所示。

【文件 7-4】 res/drawable/inverse_btn_selector.xml

```xml
<?xml version="1.0" encoding="utf-8"?>
<selector xmlns:android="http://schemas.android.com/apk/res/android" >
    <item android:state_pressed="true" android:drawable="@drawable/inverse_
    btn_p"/>
    <item android:state_pressed="false" android:drawable="@drawable/inverse_
    btn_n"/>
</selector>
```

7.2.2 进程管理 Item 布局

由于进程管理界面中 ListView 展示数据的布局都一样,因此可以定义一个 Item 条目布局用于展示数据,具体代码如【文件 7-5】所示。

【文件 7-5】 item_processmanager_list.xml

```xml
<?xml version="1.0" encoding="utf-8"?>
<RelativeLayout xmlns:android="http://schemas.android.com/apk/res/android"
    android:layout_width="match_parent"
    android:layout_height="wrap_content"
    android:orientation="vertical">
    <ImageView
        android:id="@+id/imgv_appicon_processmana"
        android:layout_width="50dp"
        android:layout_height="50dp"
        android:layout_alignParentLeft="true"
        android:layout_centerVertical="true"
        android:layout_margin="5dp"/>
    <TextView
        android:id="@+id/tv_appname_processmana"
        style="@style/textview16sp"
        android:layout_width="150dp"
        android:layout_alignParentTop="true"
        android:singleLine="true"
        android:layout_marginLeft="16dp"
        android:layout_toRightOf="@+id/imgv_appicon_processmana"
        android:layout_margin="5dp"
```

```xml
            android:textColor="@color/black" />
        <TextView
            android:id="@+id/tv_appmemory_processmana"
            style="@style/textview14sp"
            android:layout_margin="5dp"
            android:layout_below="@+id/tv_appname_processmana"
            android:layout_toRightOf="@+id/imgv_appicon_processmana"
            android:textColor="@color/black"/>
        <CheckBox
            android:id="@+id/checkbox"
            style="@style/wrapcontent"
            android:button="@drawable/green_checkbox_selector"
            android:layout_alignParentRight="true"
            android:layout_centerVertical="true"
            android:focusable="false"
            android:clickable="false"
            android:layout_marginRight="10dp"/>
</RelativeLayout>
```

上述布局文件中，ImageView 控件用于展示程序图标，第一个 TextView 控件用于展示程序名称，第二个 TextView 用于展示程序占用内存大小，CheckBox 控件用于选中当前程序。CheckBox 控件还使用了背景选择器，具体如【文件 7-6】所示。

【文件 7-6】res/drawable/green_checkbox_selector.xml

```xml
<?xml version="1.0" encoding="utf-8"?>
<selector xmlns:android="http://schemas.android.com/apk/res/android" >
    <item android:state_checked="true" android:drawable="@drawable/green_ checkbox_c"/>
    <item android:state_checked="false" android:drawable="@drawable/check box_n"/>
</selector>
```

7.2.3 进程信息的实体类

在 ListView 列表展示的进程信息中，包含应用名称、应用图标、程序包名、占用内存大小、是否被选中、是否为用户程序等，因此需要定义一个进程的实体类来封装这些信息，具体代码如【文件 7-7】所示。

【文件 7-7】TaskInfo.java

```java
/**正在运行的 App 的信息*/
public class TaskInfo{
    public String appName;
```

```
    public Drawable appIcon;
    public String  packageName;
    public long appMemory;
    /**用来标记app是否被选中*/
    public boolean isChecked;
    public boolean isUserApp;
}
```

7.2.4 主界面逻辑

进程管理界面主要是将所有正在运行的进程分为用户进程和系统进程展示出来，并且当选中某些进程时，点击"清理"按钮可以将进程清理掉，具体代码如【文件7-8】所示。

【文件7-8】ProcessManagerActivity.java

```
1  public class ProcessManagerActivity extends Activity implements OnClickListener{
2      private TextView mRunProcessNum;
3      private TextView mMemoryTV;
4      private TextView mProcessNumTV;
5      private ListView mListView;
6      ProcessManagerAdapter adapter;
7      private List<TaskInfo> runningTaskInfos;
8      private List<TaskInfo> userTaskInfos=new ArrayList<TaskInfo>();
9      private List<TaskInfo> sysTaskInfo=new ArrayList<TaskInfo>();
10     private ActivityManager manager;
11     private int runningPocessCount;
12     private long totalMem;
13     @Override
14     protected void onCreate(Bundle savedInstanceState){
15         super.onCreate(savedInstanceState);
16         requestWindowFeature(Window.FEATURE_NO_TITLE);
17         setContentView(R.layout.activity_processmanager);
18         initView();
19         fillData();
20     }
21     @Override
22     protected void onResume(){
23         if(adapter!=null){
24             adapter.notifyDataSetChanged();
25         }
26         super.onResume();
27     }
```

```java
28  private void initView(){
29      findViewById(R.id.rl_titlebar).setBackgroundColor(
30      getResources().getColor(R.color.bright_green));
31      ImageView mLeftImgv = (ImageView) findViewById(R.id.imgv_leftbtn);
32      mLeftImgv.setOnClickListener(this);
33      mLeftImgv.setImageResource(R.drawable.back);
34      ImageView mRightImgv = (ImageView) findViewById(R.id.imgv_rightbtn);
35      mRightImgv.setImageResource(R.drawable.processmanager_setting_icon);
36      mRightImgv.setOnClickListener(this);
37      ((TextView) findViewById(R.id.tv_title)).setText("进程管理");
38      mRunProcessNum=(TextView) findViewById(R.id.tv_runningprocess_num);
39      mMemoryTV = (TextView) findViewById(R.id.tv_memory_processmanager);
40      mProcessNumTV=(TextView) findViewById(R.id.tv_user_runningprocess);
41      runningPocessCount=SystemInfoUtils.
42      getRunningPocessCount(ProcessManagerActivity.this);
43      mRunProcessNum.setText("运行中的进程: "+runningPocessCount+ "个");
44      long totalAvailMem=SystemInfoUtils.getAvailMem(this);
45      totalMem=SystemInfoUtils.getTotalMem();
46      mMemoryTV.setText("可用/总内存: "+
47      Formatter.formatFileSize(this, totalAvailMem)+"/"+
48      Formatter.formatFileSize(this, totalMem));
49      mListView=(ListView) findViewById(R.id.lv_runningapps);
50      initListener();
51  }
52  private void initListener(){
53      findViewById(R.id.btn_selectall).setOnClickListener(this);
54      findViewById(R.id.btn_select_inverse).setOnClickListener(this);
55      findViewById(R.id.btn_cleanprocess).setOnClickListener(this);
56      mListView.setOnItemClickListener(new OnItemClickListener(){
57          @Override
58          public void onItemClick(AdapterView<?> parent,View view,
59          int position,long id){
60              Object object=mListView.getItemAtPosition(position);
61              if(object!=null & object instanceof TaskInfo){
62                  TaskInfo info=(TaskInfo) object;
63                  if(info.packageName.equals(getPackageName())){
64                      //当前点击的条目是本应用程序
65                      return;
66                  }
```

```
67              info.isChecked=!info.isChecked;
68              adapter.notifyDataSetChanged();
69           }
70        }
71     });
72     mListView.setOnScrollListener(new OnScrollListener(){
73        @Override
74        public void onScrollStateChanged(AbsListView view, int scrollState){
75        }
76        @Override
77        public void onScroll(AbsListView view, int firstVisibleItem,
78        int visibleItemCount,int totalItemCount){
79           if(firstVisibleItem>=userTaskInfos.size()+1){
80              mProcessNumTV.setText("系统进程: "+sysTaskInfo.size()+"个");
81           }else{
82              mProcessNumTV.setText("用户进程:  "+userTaskInfos.size()+"个");
83           }
84        }
85     });
86  }
87  private void fillData(){
88     userTaskInfos.clear();
89     sysTaskInfo.clear();
90     new Thread(){
91        public void run(){
92           runningTaskInfos=TaskInfoParser.
93           getRunningTaskInfos(getApplicationContext());
94           runOnUiThread(new Runnable(){
95              @Override
96              public void run(){
97                 for(TaskInfo taskInfo:runningTaskInfos){
98                    if(taskInfo.isUserApp){
99                       userTaskInfos.add(taskInfo);
100                   }else{
101                      sysTaskInfo.add(taskInfo);
102                   }
103                }
104                if(adapter==null){
105                   adapter=new ProcessManagerAdapter(
```

```
106                    getApplicationContext(),userTaskInfos,sysTaskInfo);
107                    mListView.setAdapter(adapter);
108                }else{
109                    adapter.notifyDataSetChanged();
110                }
111                if(userTaskInfos.size()>0){
112                    mProcessNumTV.setText("用户进程: "+
113                    userTaskInfos.size()+"个");
114                }else{
115                    mProcessNumTV.setText("系统进程: "+sysTaskInfo.size()+
116                    "个");
117                }
118            }
119        });
120    };
121    }.start();
122    }
123    @Override
124    public void onClick(View v){
125       switch (v.getId()){
126       case R.id.imgv_leftbtn:
127           finish();
128           break;
129       case R.id.imgv_rightbtn:
130           //跳转至进程管理设置页面
131           startActivity(new Intent(this,ProcessManagerSettingActivity.class));
132           break;
133       case R.id.btn_selectall:
134           selectAll();
135           break;
136       case R.id.btn_select_inverse:
137           inverse();
138           break;
139       case R.id.btn_cleanprocess:
140           cleanProcess();
141           break;
142       }
143    }
144    /**
```

```
145      * 清理进程
146      */
147     private void cleanProcess(){
148         manager=(ActivityManager) getSystemService(ACTIVITY_SERVICE);
149         int count=0;
150         long saveMemory=0;
151         List<TaskInfo> killedtaskInfos=new ArrayList<TaskInfo>();
152         //注意,遍历集合时不能改变集合大小
153         for(TaskInfo info:userTaskInfos){
154            if(info.isChecked){
155                count++;
156                saveMemory+=info.appMemory;
157                manager.killBackgroundProcesses(info.packageName);
158                killedtaskInfos.add(info);
159            }
160         }
161         for(TaskInfo info:sysTaskInfo){
162            if(info.isChecked){
163                count++;
164                saveMemory+=info.appMemory;
165                manager.killBackgroundProcesses(info.packageName);
166                killedtaskInfos.add(info);
167            }
168         }
169         for(TaskInfo info:killedtaskInfos){
170            if(info.isUserApp){
171                userTaskInfos.remove(info);
172            }else
173                sysTaskInfo.remove(info);
174            }
175         }
176         runningPocessCount-=count;
177         mRunProcessNum.setText("运行中的进程: "+runningPocessCount+"个");
178         mMemoryTV.setText("可用/总内存: "+Formatter.formatFileSize(this,
179         SystemInfoUtils.getAvailMem(this))+"/"+
180         Formatter.formatFileSize(this, totalMem));
181         Toast.makeText(this,"清理了"+count + "个进程,释放了"+
182         Formatter.formatFileSize(this, saveMemory)+"内存",1).show();
183         mProcessNumTV.setText("用户进程: "+userTaskInfos.size()+"个");
```

```
184        adapter.notifyDataSetChanged();
185    }
186    /**
187     * 反选
188     */
189    private void inverse(){
190        for (TaskInfo taskInfo:userTaskInfos){
191            //就是本应用程序
192            if(taskInfo.packageName.equals(getPackageName())){
193                continue;
194            }
195            boolean checked=taskInfo.isChecked;
196            taskInfo.isChecked=!checked;
197        }
198        for(TaskInfo taskInfo:sysTaskInfo){
199            boolean checked=taskInfo.isChecked;
200            taskInfo.isChecked=!checked;
201        }
202        adapter.notifyDataSetChanged();
203    }
204    /**
205     * 全选
206     */
207    private void selectAll(){
208        for(TaskInfo taskInfo : userTaskInfos){
209            //就是本应用程序
210            if(taskInfo.packageName.equals(getPackageName())){
211                continue;
212            }
213            taskInfo.isChecked=true;
214        }
215        for(TaskInfo taskInfo:sysTaskInfo){
216            taskInfo.isChecked=true;
217        }
218        adapter.notifyDataSetChanged();
219    }
220 }
```

代码说明：

- 第 22~27 行的 onResume() 方法是当 Activity 可见时，如果 ListView 的适配器对象存在，则直接调用 Adapter 的 notifyDataSetChanged() 方法更新已生成的数据列表（适配器中的数据）。
- 第 28~51 行的 initView() 方法用于初始化控件，其中 41~48 行代码通过调用 SystemInfoUtils 类，获取正在运行的进程个数、可用内存空间、可用/总内存大小。其中 SystemInfoUtils 类是一个工具类，在 7.3.1 小节会进行详细讲解。
- 第 52~86 行的 initListener() 方法用于初始化界面，给按钮添加点击监听，以及设置 ListView 的条目点击事件和滚动事件。在 ListView 的条目点击监听的 onItemClick() 方法中，循环遍历所有正在运行的程序，然后通过 TaskInfo 的 isChecked 属性标记是否被选中，并更新数据。在 ListView 滚动事件监听的 onScroll() 方法中，用于判断程序在滚动的过程中 mProcessNumTV 文本显示的内容，当滚动的数目超出用户进程个数时，显示系统进程的个数，否则显示用户进程个数。
- 第 87~122 行的 fillData() 方法用于填充数据，在子线程中通过 TaskInfoParser 类的 getRunningTaskInfos()（后面进行详细讲解）方法获取所有正在运行的程序，并且将用户程序和系统程序分别存储在 userTaskInfos 和 sysTaskInfo 集合中。然后判断 Adapter 是否为 null，如果为 null，则创建适配器对象填充 ListView 中的数据，否则直接调用 notifyDataSetChanged() 方法刷新 ListView 界面。最后通过 if 语句判断 mProcessNumTV 文本标签的显示内容，如果用户集合大于 0，则显示用户进程个数，否则显示系统进程个数。
- 第 147~185 行的 cleanProcess() 方法用于清理进程，首先分别通过 for 循环遍历用户进程和系统进程，如果进程处于被选中状态，那就将其清理，并记录清理的个数添加到 killedtaskInfos 集合（清理进程的集合）中。然后遍历 killedtaskInfos 集合，并判断是用户进程还是系统进程，然后将其从对应的集合中移除，最后展示运行中的进程个数，以及可用/总内存大小，清理了多少进程以及释放了多少内存。
- 第 189~203 行的 inverse() 方法用于控制反选按钮，首先循环遍历用户程序，并判断当前程序是否为本应用程序，然后定义 boolean 类型变量，记录当前程序是否被选中，如果被选中则将 taskInfo.isChecked 状态设置为 !checked 未被选中状态，以此来实现反选功能，同理实现系统程序的反选逻辑也是一样的。
- 第 207~219 行的 selectAll() 方法用于控制全选按钮，同样是分别遍历用户程序和系统程序，然后将 taskInfo.isChecked 属性设置为 true 即可，表示全部选中。

7.2.5 数据适配器

在进程管理界面中，使用了 ListView 控件展示所有的进程列表，因此需要定义一个数据适配器为 ListView 控件填充数据，具体代码如【文件 7-9】所示。

【文件 7-9】 ProcessManagerAdapter.java

```
1  public class ProcessManagerAdapter extends BaseAdapter{
2      private Context context;
3      private List<TaskInfo> mUsertaskInfos;
4      private List<TaskInfo> mSystaskInfos;
```

```java
5   private SharedPreferences mSP;
6   public ProcessManagerAdapter(Context context,List<TaskInfo> userTaskInfos,
7   List<TaskInfo> sysTaskInfo){
8       super();
9       this.context=context;
10      this.mUsertaskInfos=userTaskInfos;
11      this.mSystaskInfos=sysTaskInfo;
12      mSP=context.getSharedPreferences("config",Context.MODE_PRIVATE);
13  }
14  @Override
15  public int getCount(){
16      if(mSystaskInfos.size()>0 & mSP.getBoolean("showSystemProcess",true)){
17          return mUsertaskInfos.size()+mSystaskInfos.size()+2;
18      }else{
19          return mUsertaskInfos.size()+1;
20      }
21  }
22  @Override
23  public Object getItem(int position){
24      if(position==0||position==mUsertaskInfos.size()+1){
25          return null;
26      }else if(position<=mUsertaskInfos.size()){
27          //用户进程
28          return mUsertaskInfos.get(position-1);
29      }else{
30          //系统进程
31          return mSystaskInfos.get(position-mUsertaskInfos.size()-2);
32      }
33  }
34  @Override
35  public long getItemId(int position){
36      return position;
37  }
38  @Override
39  public View getView(int position, View convertView,ViewGroup parent){
40      if(position==0){
41          TextView tv=getTextView();
42          tv.setText("用户进程 : "+mUsertaskInfos.size()+"个");
43          return tv;
```

```java
44      }else if(position==mUsertaskInfos.size()+1){
45          TextView tv=getTextView();
46          if(mSystaskInfos.size()>0){
47              tv.setText("系统进程: "+mSystaskInfos.size()+"个");
48              return tv;
49          }
50      }
51      //获取TaskInfo对象
52      TaskInfo taskInfo=null;
53      if(position<=mUsertaskInfos.size()){
54          taskInfo=mUsertaskInfos.get(position-1);
55      }else if(mSystaskInfos.size()>0){
56          taskInfo=mSystaskInfos.get(position-mUsertaskInfos.size()-2);
57      }
58      ViewHolder holder=null;
59      if(convertView!=null && convertView instanceof RelativeLayout){
60          holder=(ViewHolder) convertView.getTag();
61      }else{
62          convertView=View.inflate(context,
63              R.layout.item_processmanager_list,null);
64          holder=new ViewHolder();
65          holder.mAppIconImgv=(ImageView) convertView.findViewById(
66              R.id.imgv_appicon_processmana);
67          holder.mAppMemoryTV=(TextView) convertView.findViewById(
68              R.id.tv_appmemory_processmana);
69          holder.mAppNameTV=(TextView) convertView.findViewById(
70              R.id.tv_appname_processmana);
71          holder.mCheckBox=(CheckBox) convertView.findViewById(
72              R.id.checkbox);
73          convertView.setTag(holder);
74      }
75      if(taskInfo!=null){
76          holder.mAppNameTV.setText(taskInfo.appName);
77          holder.mAppMemoryTV.setText("占用内存: "+Formatter.
78              formatFileSize(context,taskInfo.appMemory));
79          holder.mAppIconImgv.setImageDrawable(taskInfo.appIcon);
80          if(taskInfo.packageName.equals(context.getPackageName())){
81              holder.mCheckBox.setVisibility(View.GONE);
82          }else{
```

```
83              holder.mCheckBox.setVisibility(View.VISIBLE);
84          }
85          holder.mCheckBox.setChecked(taskInfo.isChecked);
86      }
87      return convertView;
88  }
89  /**
90   * 创建一个 TextView
91   * @return
92   */
93  private TextView getTextView(){
94      TextView tv=new TextView(context);
95      tv.setBackgroundColor(context.getResources().getColor(color.graye5));
96      tv.setPadding(DensityUtil.dip2px(context,5),DensityUtil.dip2px
97      (context, 5),DensityUtil.dip2px(context,5),
98      DensityUtil.dip2px(context, 5));
99      tv.setTextColor(context.getResources().getColor(R.color.black));
100     return tv;
101 }
102 static class ViewHolder{
103     ImageView mAppIconImgv;
104     TextView mAppNameTV;
105     TextView mAppMemoryTV;
106     CheckBox mCheckBox;
107 }
108 }
```

代码说明：

- 第 15~21 行的 getCount()方法用于返回总条目数，如果系统程序大于 0 并且 SharedPreferences 中存储的 key 为 showSystemProcess 的值为 true 时，会返回"用户程序+系统程序+2"，这个 2 表示的是 mProcessNumTV 的数量（展示是用户程序还是系统程序的文本标签）。否则返回"用户程序+1"，这个 1 表示展示用户程序个数的 mProcessNumTV 文本标签。

- 第 23~33 行的 getItem()方法用于返回当前进程对象，首先判断当前位置是否为用户程序个数和系统程序个数的文本标签，如果不是则判断是否为用户进程，如果是则返回用户进程对象（mUsertaskInfos.get(position −1)将用户进程标签减掉），否则返回系统进程对象（mSystaskInfos.get(position−mUsertaskInfos.size() −2);将用户进程个数和两个文本标签都减掉）。

- 第 93～101 行的 getTextView()方法用于创建一个 TextView，该控件在第 41 行和第 45 行代码处使用，分别用于显示用户进程个数、系统进程个数。其中用到的 DensityUtil 对象是一个单位转换的工具类，前面已经讲解过，直接使用即可。

7.3 工 具 类

7.3.1 获取系统信息

在主界面逻辑代码中，使用了 SystemInfoUtils 工具类，该类主要用于获取系统信息，它有四个方法，分别用于判断一个服务是否处于运行状态（设置界面逻辑代码中调用）、获取手机总内存大小、获取可用的内存信息、得到正在运行进程的数量，具体代码如【文件 7-10】所示。

【文件 7-10】SystemInfoUtils.java

```
1  public class SystemInfoUtils{
2      /**
3       * 判断一个服务是否处于运行状态
4       * @param context 上下文
5       * @return
6       */
7      public static boolean isServiceRunning(Context context,String className){
8          ActivityManager am=(ActivityManager) context.
9              getSystemService(Context.ACTIVITY_SERVICE);
10         List<RunningServiceInfo> infos=am.getRunningServices(200);
11         for(RunningServiceInfo info:infos){
12             String serviceClassName=info.service.getClassName();
13             if(className.equals(serviceClassName)){
14                 return true;
15             }
16         }
17         return false;
18     }
19     /**
20      * 获取手机的总内存大小单位byte
21      * @return
22      */
23     public static long getTotalMem(){
24         try{
25             FileInputStream fis=new FileInputStream(new
```

```
26          File("/proc/ meminfo"));
27          BufferedReader br=new BufferedReader(new InputStreamReader(fis));
28          String totalInfo=br.readLine();
29          StringBuffer sb=new StringBuffer();
30          for(char c:totalInfo.toCharArray()){
31           if(c>='0' && c<='9'){
32              sb.append(c);
33           }
34          }
35          long bytesize=Long.parseLong(sb.toString())*1024;
36          return bytesize;
37       } catch (Exception e){
38          e.printStackTrace();
39          return 0;
40       }
41   }
42   /**
43    * 获取可用的内存信息
44    * @param context
45    * @return
46    */
47   public static long getAvailMem(Context context){
48      ActivityManager am=(ActivityManager)
49      context.getSystemService(Context.ACTIVITY_SERVICE);
50      //获取内存大小
51      MemoryInfo outInfo=new ActivityManager.MemoryInfo();
52      am.getMemoryInfo(outInfo);
53      long availMem=outInfo.availMem;
54      return availMem;
55   }
56   /**
57    * 得到正在运行的进程的数量
58    * @param context
59    * @return
60    */
61   public static int getRunningPocessCount(Context context){
62      ActivityManager am=(ActivityManager)
63      context.getSystemService(Context.ACTIVITY_SERVICE);
```

```
64      List<RunningAppProcessInfo> runningAppProcessInfos=
65      am.getRunningAppProcesses();
66      int count=runningAppProcessInfos.size();
67      return count;
68   }
69 }
```

7.3.2 获取进程信息

在进程管理界面中,需要获取正在运行的进程,该功能代码相对比较独立,因此将其抽取出来作为一个单独的工具类供其使用,具体代码如【文件 7-11】所示。

【文件 7-11】 TaskInfoParser.java

```
1  /**
2   * 任务信息 & 进程信息的解析器
3   * @author Administrator
4   */
5  public class TaskInfoParser{
6      /**
7       * 获取正在运行的所有进程的信息
8       * @param context 上下文
9       * @return 进程信息的集合
10      */
11     public static List<TaskInfo> getRunningTaskInfos(Context context){
12         ActivityManager am=(ActivityManager) context.
13         getSystemService(Context.ACTIVITY_SERVICE);
14         PackageManager pm=context.getPackageManager();
15         List<RunningAppProcessInfo> processInfos = am.getRunningAppProcesses();
16         List<TaskInfo> taskInfos=new ArrayList<TaskInfo>();
17         for(RunningAppProcessInfo processInfo:processInfos){
18             String packname=processInfo.processName;
19             TaskInfo taskInfo=new TaskInfo();
20             taskInfo.packageName=packname;//进程名称
21             MemoryInfo[] memroyinfos=am.getProcessMemoryInfo(
22             new int[]{processInfo.pid});
23             long memsize=memroyinfos[0].getTotalPrivateDirty()*1024;
24             taskInfo.appMemory=memsize;//程序占用的内存空间
25             try{
```

```
26            PackageInfo packInfo=pm.getPackageInfo(packname, 0);
27            Drawable icon=packInfo.applicationInfo.loadIcon(pm);
28            taskInfo.appIcon=icon;
29            String appname=packInfo.applicationInfo.loadLabel(pm).toString();
30            taskInfo.appName=appname;
31            if((ApplicationInfo.FLAG_SYSTEM&packInfo.applicationInfo.
32            flags)!=0){
33                //系统进程
34                taskInfo.isUserApp=false;
35            }else{
36                //用户进程
37                taskInfo.isUserApp=true;
38            }
39        } catch(NameNotFoundException e){
40            e.printStackTrace();
41            taskInfo.appName=packname;
42            taskInfo.appIcon=
43            context.getResources().getDrawable(R.drawable.ic_default);
44        }
45        taskInfos.add(taskInfo);
46    }
47    return taskInfos;
48 }
49}
```

代码说明：

- 第 14～15 行代码用于获取一个包管理器 PackageManager，通过包管理器获取正在运行的程序。
- 第 16 行代码创建了一个 List<TaskInfo> 进程集合，用于存储进程对象。
- 第 17～46 行代码用于循环遍历所有正在运行的进程，并获取程序包名、占用内存大小、程序图标、程序名称，以及是否是用户程序，最后将进程信息封装到 TaskInfo 对象中并添加到 List<TaskInfo> 集合。

7.4 设 置 进 程

在清理进程时，通常不希望手动进行清理，有时还希望在手机睡眠时自动清理进程，因此可以利用服务和广播接收者在锁屏时自动清理进程。本节将针对如何控制是否显示系统进程以及锁屏时是否清理进程功能进行详细讲解。

7.4.1 设置进程界面

设置进程界面主要由两个相对布局和两个 View 对象组成，每个相对布局中放置一个 ImageView 控件、TextView 控件、ToggleButton 控件，分别用于显示功能图标、功能文字、切换按钮，设置进程的图形化界面如图 7-4 所示。

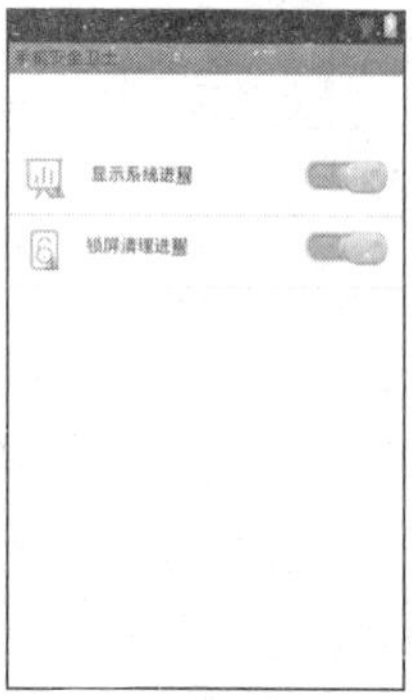

图 7-4 所示设置进程界面对应的布局如【文件 7-12】所示。

【文件 7-12】 activity_processmanagersetting.xml

图 7-4 进程设置界面

```xml
<?xml version="1.0" encoding="utf-8"?>
<LinearLayout xmlns:android="http://schemas.android.
   com/apk/res/android"
    android:layout_width="match_parent"
    android:layout_height="match_parent"
    android:orientation="vertical">
    <include layout="@layout/titlebar"/>
    <RelativeLayout
        android:layout_width="match_parent"
        android:layout_height="55dp"
        android:gravity="center_vertical">
        <ImageView
            android:id="@+id/imageView1"
            style="@style/wrapcontent"
            android:layout_alignParentLeft="true"
            android:layout_marginLeft="10dp"
            android:background="@drawable/processmanager_showsysapps"/>
        <TextView
            style="@style/wrapcontent"
            android:layout_marginLeft="20dp"
            android:layout_marginTop="5dp"
            android:layout_toRightOf="@+id/imageView1"
            android:text="显示系统进程"/>
        <ToggleButton
            android:id="@+id/tgb_showsys_process"
            android:layout_width="70dp"
            android:layout_height="30dp"
            android:layout_alignParentRight="true"
            android:layout_marginRight="10dp"
            android:background="@drawable/toggle_btn_green_selector"
            android:textOff=""
```

```xml
            android:textOn=""/>
    </RelativeLayout>
    <View
        android:layout_width="match_parent"
        android:layout_height="1px"
        android:background="@color/black30"/>
    <RelativeLayout
        android:layout_width="match_parent"
        android:layout_height="55dp"
        android:gravity="center_vertical" >
        <ImageView
            android:id="@+id/imageView2"
            style="@style/wrapcontent"
            android:layout_alignParentLeft="true"
            android:layout_marginLeft="15dp"
            android:background="@drawable/processmanagerlockscreen_icon"/>
        <TextView
            style="@style/wrapcontent"
            android:layout_marginLeft="20dp"
            android:layout_marginTop="5dp"
            android:layout_toRightOf="@+id/imageView2"
            android:text="锁屏清理进程"/>
        <ToggleButton
            android:id="@+id/tgb_killprocess_lockscreen"
            android:layout_width="70dp"
            android:layout_height="30dp"
            android:layout_alignParentRight="true"
            android:layout_marginRight="10dp"
            android:background="@drawable/toggle_btn_green_selector"
            android:textOff=""
            android:textOn=""/>
    </RelativeLayout>
    <View
        android:layout_width="match_parent"
        android:layout_height="1px"
        android:background="@color/black30"/>
</LinearLayout>
```

上述布局文件中，ToggleButton 使用了一个背景选择器，用于控制按钮选中与未选中展示的背景图片,当按钮选中时,展示的是 green_troggle_c.png 图片(绿色),当按钮未选中时,展示的是 green_troggle_off.png 图片(灰色),具体代码如【文件 7-13】所示。

【文件 7-13】res/drawable/toggle_btn_green_selector.xml

```xml
<?xml version="1.0" encoding="utf-8"?>
<selector xmlns:android="http://schemas.android.com/apk/res/android" >
    <item android:state_checked="true" android:drawable="@drawable/green_ troggle_c"/>
    <item android:state_checked="false" android:drawable="@drawable/switch_ btn_off"/>
</selector>
```

7.4.2 设置进程逻辑

设置进程界面的功能是控制系统进程是否显示、锁屏时是否清理进程。当开启显示系统进程时,在进程管理界面中才会同时看到系统进程和用户进程,具体代码如【文件 7-14】所示。

【文件 7-14】ProcessManagerSettingActivity.java

```
1  public class ProcessManagerSettingActivity extends Activity implements
2  OnClickListener, OnCheckedChangeListener{
3    private ToggleButton mShowSysAppsTgb;
4    private ToggleButton mKillProcessTgb;
5    private SharedPreferences mSP;
6    private boolean running;
7    @Override
8    protected void onCreate(Bundle savedInstanceState){
9      super.onCreate(savedInstanceState);
10     requestWindowFeature(Window.FEATURE_NO_TITLE);
11     setContentView(R.layout.activity_processmanagersetting);
12     mSP=getSharedPreferences("config",MODE_PRIVATE);
13     initView();
14   }
15   /**初始化控件*/
16   private void initView(){
17     findViewById(R.id.rl_titlebar).setBackgroundColor(
18     getResources().getColor(R.color.bright_green));
19     ImageView mLeftImgv=(ImageView) findViewById(R.id.imgv_leftbtn);
20     mLeftImgv.setOnClickListener(this);
21     mLeftImgv.setImageResource(R.drawable.back);
22     ((TextView) findViewById(R.id.tv_title)).setText("进程管理设置");
23     mShowSysAppsTgb=(ToggleButton) findViewById
24     (R.id.tgb_showsys_process);
```

```java
25      mKillProcessTgb=(ToggleButton) findViewById
26          (R.id.tgb_killprocess_lockscreen);
27      mShowSysAppsTgb.setChecked(mSP.getBoolean("showSystemProcess",true));
28      running=SystemInfoUtils.isServiceRunning(this,
29          "cn.itcast.mobliesafe.chapter07.service.AutoKillProcessService");
30      mKillProcessTgb.setChecked(running);
31      initListener();
32  }
33  /**初始化监听*/
34  private void initListener(){
35      mKillProcessTgb.setOnCheckedChangeListener(this);
36      mShowSysAppsTgb.setOnCheckedChangeListener(this);
37  }
38  @Override
39  public void onClick(View v){
40      switch (v.getId()){
41      case R.id.imgv_leftbtn:
42          finish();
43          break;
44      }
45  }
46  @Override
47  public void onCheckedChanged(CompoundButton buttonView, boolean isChecked){
48      switch (buttonView.getId()){
49      case R.id.tgb_showsys_process:
50          saveStatus("showSystemProcess",isChecked);
51          break;
52      case R.id.tgb_killprocess_lockscreen:
53          Intent service=new Intent(this,AutoKillProcessService.class);
54          if(isChecked){
55              //开启服务
56              startService(service);
57          }else{
58              //关闭服务
59              stopService(service);
60          }
61          break;
62      }
63  }
64  private void saveStatus(String string, boolean isChecked){
```

```
65        Editor edit=mSP.edit();
66        edit.putBoolean(string,isChecked);
67        edit.commit();
68    }
69 }
```

代码说明:

- 第 16~32 行的 initView()方法用于初始化控件,其中第 27 行代码用于设置是否显示系统进程,第 28~30 行代码用于判断自动杀死进程服务是否运行,并将结果设置给 mKillProcessTgb 控件(锁屏清理进程)。
- 第 47~63 行的 onCheckedChanged()方法用于监听 ToggleButton 按钮的状态改变,当"显示系统进程"按钮状态改变时,调用 saveStatus()方法将 showSystemProcess 状态保存,当"锁屏清理进程"按钮状态是选中时时,会启动 AutoKillProcessService 服务,否则将服务停止。
- 第 64~68 行代码用于保存"显示进程"的按钮状态。

7.4.3 锁屏清理进程服务

当设置界面中的锁屏清理进程按钮开启时,就会打开进程清理的服务,在该服务中注册了监听屏幕锁屏的广播接收者,当屏幕锁屏时该广播接收到屏幕锁屏的消息后会自动清理进程,具体代码如【文件 7-15】所示。

【文件 7-15】AutoKillProcessService.java

```
1  public class AutoKillProcessService extends Service{
2      private ScreenLockReceiver receiver;
3      @Override
4      public IBinder onBind(Intent intent){
5          return null;
6      }
7      @Override
8      public void onCreate(){
9          super.onCreate();
10         receiver=new ScreenLockReceiver();
11         registerReceiver(receiver,new IntentFilter(Intent.ACTION_SCREEN_OFF));
12     }
13     @Override
14     public void onDestroy(){
15         unregisterReceiver(receiver);
16         receiver=null;
17         super.onDestroy();
18     }
19     class ScreenLockReceiver extends BroadcastReceiver{
```

```java
20      @Override
21      public void onReceive(Context context, Intent intent){
22          ActivityManager am=(ActivityManager)
23          getSystemService(ACTIVITY_SERVICE);
24          for(RunningAppProcessInfo info: am.getRunningAppProcesses()){
25              String packname=info.processName;
26              am.killBackgroundProcesses(packname);
27          }
28      }
29  }
30 }
```

代码说明：
- 第8～12行的onCreate()方法中动态的注册了一个广播接收者，用于接收屏幕关闭的广播。
- 第14～18行的onDestroy()方法中将广播接收者注销并设置为null。
- 第19～29行创建了一个锁屏的广播接收者ScreenLockReceiver，在该广播接收者中获取正在运行的进程，并将其杀死。

接下来在AndroidManifest.xml文件中配置清理进程服务，具体代码如下：

```xml
<!-- 锁屏自动清理进程 -->
<service
    android:name="cn.itcast.mobliesafe.chapter07.service.AutoKillProcessService"
        android:persistent="true" >
</service>
```

上述配置中，加入android:persistent="true"属性后，就可以保证该应用程序所在进程不会被LMK（LowMemoryKiller，表示最小的LKM内存阀值，例如系统默认设置为50 MB，当手机内存小于50 MB时，LMK机制开始关闭进程）杀死。

本章小结

本章主要针对进程管理模块进行讲解，首先针对该模块功能进行介绍，然后讲解如何将进程信息展示到ListView中以及需要使用的工具类，最后讲解了如何设置是否显示系统进程以及锁屏清理进程。通过该模块的学习，编程者可以更熟练掌握如何管理系统中的进程以及杀死进程。

【面试精选】
1. 请问进程间通信有几种方式？
2. 请问启动、停止Service有几种方式，各自有什么特点？
扫描右方二维码，查看面试题答案！

第 8 章

→ 流量统计模块

学习目标
- 了解流量统计模块功能；
- 掌握数据库的创建与使用；
- 掌握如何获取流量信息。

在日常生活中，大家使用手机上网时用的基本都是 3G/4G 网络，虽然速度比 2G 网络要快很多，但也消耗了大量手机流量，并且 Android 中有些应用还会在后台偷偷联网耗费流量，而每个月手机流量都是有限的，如果超过了上限会造成一定的经济损失，因此流量使用也是大家非常关心的问题。本章将针对流量统计模块进行详细讲解。

8.1 模 块 概 述

8.1.1 功能介绍

流量统计模块的主要作用就是统计当月手机的总流量、本月已用流量以及本日已用流量信息。在使用该功能之前首先需要手动设置运营商，然后点击"完成"按钮进入流量监控界面显示本月流量使用情况，具体如图 8-1 所示。

图 8-1 流量监控界面

8.1.2 代码结构

流量统计模块代码较少，结构清晰，接下来通过一个图例来展示流量监控模块的代码结构，具体如图 8-2 所示。

图 8-2 chapter08 代码结构

下面按照结构顺序依次介绍 chapter08 包中的文件，具体如下：
- TrafficDao.java：对流量数据库进行增加、修改、查询的工具类；
- TrafficOpenHelper.java：保存流量信息的数据库；
- BootCompleteReceiver.java：监听开机启动的广播接收者，当手机开机时打开服务；
- TrafficMonitoringService.java：获取流量具体数据的服务；
- SystemInfoUtils.java：判断服务是否开启的工具类；
- OperatorSetActivity.java：信息设置界面逻辑，第一次进入时需要设置运营商信息；
- TrafficMonitoringActivity.java：流量监控界面逻辑，用于展示本月、本日已用流量和本月总流量。

8.2 运营商设置

手机中使用的流量都来自于三大运营商：联通、移动、电信。每个运营商的套餐信息是不同的，因此在统计流量信息时需要设置手机使用的是哪一个运营商，然后根据不同运营商进行不同的处理。本节将针对运营商设置进行详细讲解。

8.2.1 运营商设置界面

运营商设置界面是为了查询流量套餐使用的，在查询流量时需要根据运营商发送指定的短信命令。该界面中分为三部分，第一部分有一张提示图片和一段文字组成；第二部分为一个 TextView 控件以及一个下拉控件,用于选择运营商信息;最下方放置一个按钮,点击之后进入流量监控界面,该界面在下次进入时不会再出现。运营商设置图形化界面如图 8-3 所示。

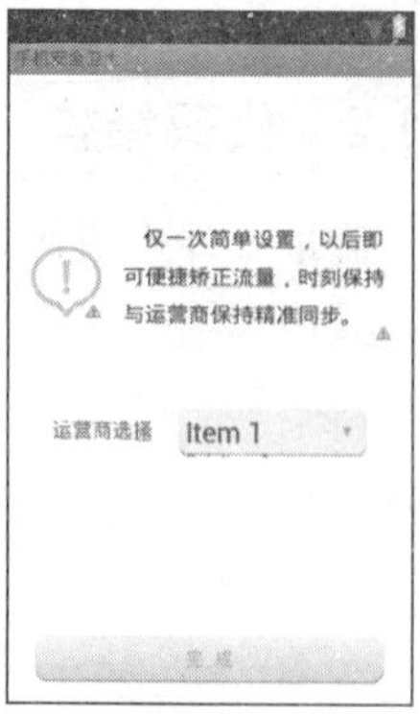

图 8-3 所示运营商设置界面对应的布局如【文件 8-1】所示。

【文件 8-1】activity_operatorset.xml

图 8-3 运营商设置界面

```xml
<?xml version="1.0" encoding="utf-8"?>
<LinearLayout xmlns:android="http://schemas.android.com/apk/res/android"
    android:layout_width="match_parent"
    android:layout_height="match_parent"
    android:orientation="vertical" >
    <LinearLayout
        android:layout_width="match_parent"
        android:layout_height="match_parent"
        android:layout_weight="100"
        android:orientation="vertical">
        <include layout="@layout/titlebar"/>
        <LinearLayout
            style="@style/wrapcontent"
            android:layout_marginLeft="15dp"
            android:layout_marginRight="15dp"
            android:layout_marginTop="60dp"
            android:gravity="center"
            android:orientation="horizontal">
            <ImageView
                android:layout_width="60dp"
                android:layout_height="60dp"
                android:background="@drawable/operator_set_info_incon"/>
            <TextView
                style="@style/textview16sp"
                android:layout_marginLeft="15dp"
                android:lineSpacingMultiplier="1.5"
                android:text="@string/_operatorset_string"
                android:textColor="@color/black"
                android:textScaleX="1.1"/>
        </LinearLayout>
        <LinearLayout
            android:layout_width="match_parent"
            android:layout_height="wrap_content"
```

```xml
            android:layout_marginTop="50dp"
            android:gravity="center"
            android:orientation="horizontal" >
            <TextView
                style="@style/textview16sp"
                android:text="运营商选择" />
            <Spinner
                android:id="@+id/spinner_operator_select"
                android:layout_width="150dp"
                android:layout_height="40dp"
                android:layout_marginLeft="20dp"
                android:spinnerMode="dropdown"
                android:background="@drawable/operator_spinner_bg"
                android:gravity="center" />
        </LinearLayout>
    </LinearLayout>
    <Button
        android:id="@+id/btn_operator_finish"
        android:layout_weight="10"
        android:layout_width="280dp"
        android:layout_height="45dp"
        android:layout_gravity="center_horizontal"
        android:background="@drawable/operator_finish_bg_selector"
        android:layout_marginBottom="15dp"/>
</LinearLayout>
```

在上述代码中，使用了两个自定义样式 style="@style/wrapcontent"和 style= "@style/textview16sp"，指定控件的宽高都为包裹内容，其中 style="@style/textview16sp"继承自 style="@style/wrapcontent"，在指定控件的宽高基础上，还指定了文字大小和颜色，具体请查阅 styles.xml 文件。

由于运营商设置界面中使用了一个 Spinner 控件，该控件与 ListView 控件类似需要设置 Item 中的数据，因此，需要定义一 item_spinner_operatorset.xml 布局放置一个 TextView 控件，用于显示运营商选项，具体如【文件 8-2】所示。

【文件 8-2】item_spinner_operatorset.xml

```xml
<?xml version="1.0" encoding="utf-8"?>
<LinearLayout xmlns:android="http://schemas.android.com/apk/res/android"
    android:layout_width="match_parent"
    android:layout_height="match_parent"
    android:orientation="vertical" >
    <TextView
```

```
        android:id="@+id/tv_provice"
        style="@style/textview16sp"
        android:layout_margin="8dp"/>
</LinearLayout>
```

最下方的 Button 按钮使用了图片选择器 operator_finish_bg_selector，用于控制按钮按下与弹起时的图片，当按钮按下时显示 operator_finish_p.png（绿色）图片，当按钮弹起时显示 operator_finish_n.png（灰色）图片，具体如【文件 8-3】所示。

【文件 8-3】res/drawable/operator_finish_bg_selector.xml

```xml
<?xml version="1.0" encoding="utf-8"?>
<selector xmlns:android="http://schemas.android.com/apk/res/android" >
    <item android:state_pressed="true" android:drawable="@drawable/operator_finish_p"/>
    <item android:state_pressed="false" android:drawable="@drawable/operator_finish_n"/>
</selector>
```

8.2.2 运营商设置逻辑

运营商设置界面主要是选择手机使用的运营商，并将该运营商存储到 SharedPreferences 对象中，点击确定按钮后进入流量监控界面，具体代码如【文件 8-4】所示。

【文件 8-4】OperatorSetActivity.java

```java
1  public class OperatorSetActivity extends Activity implements OnClickListener {
2      private Spinner mSelectSP;
3      private String[] operators={"中国移动","中国联通","中国电信"};
4      private ArrayAdapter mSelectadapter;
5      private SharedPreferences msp;
6      @Override
7      protected void onCreate(Bundle savedInstanceState){
8          super.onCreate(savedInstanceState);
9          requestWindowFeature(Window.FEATURE_NO_TITLE);
10         setContentView(R.layout.activity_operatorset);
11         msp = getSharedPreferences("config",MODE_PRIVATE);
12         initView();
13     }
14     @SuppressWarnings("unchecked")
15     private void initView(){
16         findViewById(R.id.rl_titlebar).setBackgroundColor(
17             getResources().getColor(R.color.light_green));
18         ImageView mLeftImgv=(ImageView) findViewById(R.id.imgv_leftbtn);
19         ((TextView) findViewById(R.id.tv_title)).setText("运营商信息设置");
20         mLeftImgv.setOnClickListener(this);
```

```
21      mLeftImgv.setImageResource(R.drawable.back);
22      mSelectSP=(Spinner) findViewById(R.id.spinner_operator_select);
23      mSelectadapter=new ArrayAdapter(this,
24      R.layout.item_spinner_operatorset, R.id.tv_provice, operators);
25      mSelectSP.setAdapter(mSelectadapter);
26      findViewById(R.id.btn_operator_finish).setOnClickListener(this);
27  }
28  @Override
29  public void onClick(View v){
30      Editor edit=msp.edit();
31      switch (v.getId()){
32      case R.id.imgv_leftbtn:
33          edit.putBoolean("isset_operator",false);
34          finish();
35          break;
36      case R.id.btn_operator_finish:
37          edit.putInt("operator",mSelectSP.getSelectedItemPosition()+1);
38          edit.putBoolean("isset_operator",true);
39          edit.commit();
40          startActivity(new Intent(this, TrafficMonitoringActivity.class));
41          finish();
42          break;
43      }
44  }
45 }
```

代码说明：

- 第 15～27 行的 initView()方法初始化界面布局控件，将运营商的信息填充到 Spinner 控件中显示在界面上。
- 第 29～44 行的 onClick()方法为按钮的点击事件，当运营商选择完毕之后点击"完成"按钮，首先会使用 SharedPreferences 将选择的运营商信息存储，并存储一个 boolean 值类型数据 isset_operator 为 true,然后会进入该模块的主界面 TrafficMonitoringActivity，在主界面中会判断 isset_operator 的值，如果该值为 true,说明运营商信息已经设置过，不需要进入运营商信息设置界面，而直接进入流量监控界面。

8.3 数据库操作

在 Android 系统中，存储每日流量可以使用 SQLite 数据库。由于向运营商发送短信只能获取到本月使用总流量和本月已用流量，而无法得到每日使用流量，因此需要自己实时进行计算，并根据日期将使用的流量存储到数据库中，然后不断更新数据库，本节将针对流量数据库操作进行详细讲解。

8.3.1 创建数据库

在进行流量统计之前，需要先创建数据库，用于存储程序所使用的流量，创建数据库代码如【文件 8-5】所示。

【文件 8-5】TrafficOpenHelper.java

```
1  public class TrafficOpenHelper extends SQLiteOpenHelper{
2      private static final String DB_NAME="traffic.db";
3      private static final String TABLE_NAME="traffic";
4      /** 流量 */
5      private final static String GPRS="gprs";
6      private final static String TIME="date";
7      public TrafficOpenHelper(Context context){
8          super(context,DB_NAME,null,1);
9      }
10     @Override
11     public void onCreate(SQLiteDatabase db){
12         db.execSQL("create table "+TABLE_NAME+"(id integer primary key
13         autoincrement,"+GPRS+"varchar(255),"+TIME+" datetime)");
14     }
15     @Override
16     public void onUpgrade(SQLiteDatabase db, int oldVersion, int newVersion) {
17     }
18 }
```

代码说明：

- 第 7~9 行代码创建了一个数据库，命名为 traffic.db。
- 第 11~14 行代码创建一张表 traffic，该表有两个属性 gprs 和 date，其中 gprs 为流量，date 为存储流量的日期。

8.3.2 数据库操作

当数据库创建完成后，需要对数据库进行操作，其中包括对流量的查询、增加、修改，具体代码如【文件 8-6】所示。

【文件 8-6】TrafficDao.java

```
1  public class TrafficDao{
2      private TrafficOpenHelper helper;
3      public TrafficDao(Context context){
4          helper=new TrafficOpenHelper(context);
5      }
6      /**
7       * 获取某一天用的流量
```

```
8      * @param dataString
9      * @return
10     */
11    public long getMoblieGPRS(String dataString) {
12        SQLiteDatabase db=helper.getReadableDatabase();
13        long gprs=0;
14        Cursor cursor=db.rawQuery("select gprs from traffic where date=?",
15        new String[]{ "datetime("+dataString+")"});
16        if(cursor.moveToNext()){
17           String gprsStr=cursor.getString(0);
18           if(!TextUtils.isEmpty(gprsStr))
19              gprs=Long.parseLong(gprsStr);
20        }else{
21           gprs=-1;
22        }
23        return gprs;
24    }
25    /**
26     * 添加今天的
27     * @param gprs
28     */
29    public void insertTodayGPRS(long gprs){
30        SQLiteDatabase db=helper.getReadableDatabase();
31        Date dNow=new Date();
32        Calendar calendar=Calendar.getInstance(); //得到日历
33        calendar.setTime(dNow);//把当前时间赋给日历
34        SimpleDateFormat sdf=new SimpleDateFormat("yyyy-MM-dd");
35        String dataString=sdf.format(dNow);
36        ContentValues values=new ContentValues();
37        values.put("gprs",String.valueOf(gprs));
38        values.put("date","datetime("+dataString+")");
39        db.insert("traffic",null,values);
40    }
41    /**
42     * 修改今天的
43     * @param gprs
44     */
45    public void updateTodayGPRS(long gprs){
46        SQLiteDatabase db=helper.getWritableDatabase();
```

```
47      Date date=new Date();
48      SimpleDateFormat sdf=new SimpleDateFormat("yyyy-MM-dd");
49      String dataString=sdf.format(date);
50      ContentValues values=new ContentValues();
51      values.put("gprs",String.valueOf(gprs));
52      values.put("date","datetime("+dataString+")");
53      db.update("traffic",values,"date=?",new String[]{"datetime("+
54      dataString+")"});
55    }
56 }
```

代码说明：

- 第 11～24 行的 getMoblieGPRS(String dataString)方法，根据接收时间进行查询某一天的流量；
- 第 29～40 行的 insertTodayGPRS(long gprs)方法，根据接收的流量信息在数据库中插入数据，所插入数据的日期为当前日期；
- 第 45～55 行的 UpdateTodayGPRS(long gprs)方法根据接收到的流量数据对当前日期的数据进行修改。

8.4 流量监控

在流量监控界面中，通过发送短信给运营商可以得到最新的本月已用流量，但用户不可能每次打开应用都发送短信去获取，而此时在界面上显示的流量信息就不是实时的，同时也不准确。为此，可以开启一个服务，实时统计已使用的流量数据，并通过 SharedPreferences 对象进行存储，这样即使用户不再发送短信校准，在界面上显示的流量数据也是最新的。本节将针对流量监控功能进行详细讲解。

8.4.1 流量监控界面

流量监控界面主要用于显示当天所用流量、本月所用流量、本月总流量。该界面包含三部分，最上方使用一个相对布局显示图片及提示信息，中间定义三个 TextView 显示流量详情，下方定义一个 Button 按钮用于校正流量，流量监控图形化界面如图 8-4 所示。

图 8-4 所示流量监控界面对应的布局文件如【文件 8-7】所示。

【文件 8-7】activity_trafficmonitoring

```
<?xml version="1.0" encoding="utf-8"?>
<LinearLayout xmlns:android="http://schemas.android.
    com/apk/res/android"
    android:layout_width="match_parent"
    android:layout_height="match_parent"
    android:orientation="vertical">
```

图 8-4 流量监控界面

```xml
<include layout="@layout/titlebar"/>
<LinearLayout
    android:layout_width="match_parent"
    android:layout_height="match_parent"
    android:layout_weight="10"
    android:orientation="vertical">
    <RelativeLayout
        android:layout_width="match_parent"
        android:layout_height="wrap_content"
        android:layout_marginTop="20dp"
        android:gravity="center_vertical">
        <ImageView
            android:id="@+id/imgv_traffic_remind"
            android:layout_width="wrap_content"
            android:layout_height="wrap_content"
            android:layout_marginLeft="30dp"
            android:src="@drawable/traffic_reminder_selector"/>
        <TextView
            android:id="@+id/tv_traffic_remind"
            style="@style/textview16sp"
            android:layout_centerVertical="true"
            android:layout_marginLeft="10dp"
            android:layout_toRightOf="@+id/imgv_traffic_remind"
            android:gravity="center"
            android:text="本月流量充足请放心使用！" />
    </RelativeLayout>
    <LinearLayout
        android:layout_width="match_parent"
        android:layout_height="wrap_content"
        android:layout_marginLeft="30dp"
        android:layout_marginRight="30dp"
        android:layout_marginTop="20dp"
        android:background="@drawable/traffic_green_bg"
        android:orientation="vertical" >
        <TextView
            android:id="@+id/tv_today_gprs"
            style="@style/textview16sp"
            android:layout_margin="5dp"
            android:text="今天已用：" />
```

```xml
        <TextView
            android:id="@+id/tv_month_usedgprs"
            style="@style/textview16sp"
            android:layout_margin="5dp"
            android:text="本月已用 : " />
        <TextView
            android:id="@+id/tv_month_totalgprs"
            style="@style/textview16sp"
            android:layout_margin="5dp"
            android:text="本月流量 : " />
        </LinearLayout>
    </LinearLayout>
    <Button
        android:id="@+id/btn_correction_flow"
        android:layout_width="280dp"
        android:layout_height="78dp"
        android:layout_gravity="center_horizontal"
        android:layout_marginBottom="10dp"
        android:layout_weight="4"
        android:background="@drawable/correction_flow_btn_selector" />
</LinearLayout>
```

在上述布局中，上方定义的 TextView 用于提示当前流量信息，剩余流量充足和流量不足时会显示不同的文字。在文字的左侧有一个 ImageView，该控件使用了图片选择器，当流量充足时显示绿色圆点，当流量不足时显示红色圆点，具体代码如【文件 8-8】所示。

【文件 8-8】res/drawable/traffic_reminder_selector.xml

```xml
<?xml version="1.0" encoding="utf-8"?>
<selector xmlns:android="http://schemas.android.com/apk/res/android" >
    <item android:state_enabled="false" android:drawable="@drawable/traffic_red"/>
    <item android:state_enabled="true" android:drawable="@drawable/traffic_green"/>
</selector>
```

除了 ImageView 按钮使用了背景选择器外，下方的"校正流量"按钮也用到了背景选择器，具体代码如【文件 8-9】所示。

【文件 8-9】res/drawable/correction_flow_btn_selector.xml

```xml
<?xml version="1.0" encoding="utf-8"?>
<selector xmlns:android="http://schemas.android.com/apk/res/android" >
    <item android:state_pressed="true" android:drawable="@drawable/correction_
```

```
        flow_p"/>
        <item android:state_pressed="false" android:drawable="@drawable/correction_
        flow_n"/>
</selector>
```

8.4.2 流量监控逻辑

在流量监控逻辑中,以联通手机为例,当点击"校正流量"按钮时,会自动向联通发送一条短信信息,获取当前流量使用情况并显示在界面上。需要注意的是,联通 3G 和 4G 所发送的流量查询短信是不一样的,接收到的短信也不一样,这里只对联通 3G 流量进行校正,联通 4G 的查询方法和 3G 大同小异,大家可以自行测试。具体代码如【文件 8-10】所示。

【文件 8-10】TrafficMonitoringActivity.java

```
1  public class TrafficMonitoringActivity extends Activity implements
2  OnClickListener{
3      private SharedPreferences mSP;
4      private Button mCorrectFlowBtn;
5      private TextView mTotalTV;
6      private TextView mUsedTV;
7      private TextView mToDayTV;
8      private TrafficDao dao;
9      private ImageView mRemindIMGV;
10     private TextView mRemindTV;
11     private CorrectFlowReceiver receiver;
12     @Override
13     protected void onCreate(Bundle savedInstanceState){
14         super.onCreate(savedInstanceState);
15         requestWindowFeature(Window.FEATURE_NO_TITLE);
16         setContentView(R.layout.activity_trafficmonitoring);
17         mSP=getSharedPreferences("config", MODE_PRIVATE);
18         boolean flag = mSP.getBoolean("isset_operator", false);
19         //如果没有设置运营商信息则进入信息设置页面
20         if(!flag){
21             startActivity(new Intent(this,OperatorSetActivity.class));
22             finish();
23         }
24         if(!SystemInfoUtils.isServiceRunning(this, "cn.itcast.mobliesafe.
25             chapter08.service.TrafficMonitoring Service")){
26             startService(new Intent(this,TrafficMonitoringService.class));
27         }
28         initView();
```

```java
29        regestReceiver();
30        initData();
31    }
32    private void initView(){
33        findViewById(R.id.rl_titlebar).setBackgroundColor(
34            getResources().getColor(R.color.light_green));
35        ImageView mLeftImgv=(ImageView) findViewById(R.id.imgv_leftbtn);
36        ((TextView) findViewById(R.id.tv_title)).setText("流量监控");
37        mLeftImgv.setOnClickListener(this);
38        mLeftImgv.setImageResource(R.drawable.back);
39        mCorrectFlowBtn=(Button) findViewById(R.id.btn_correction_flow);
40        mCorrectFlowBtn.setOnClickListener(this);
41        mTotalTV=(TextView) findViewById(R.id.tv_month_totalgprs);
42        mUsedTV=(TextView) findViewById(R.id.tv_month_usedgprs);
43        mToDayTV=(TextView) findViewById(R.id.tv_today_gprs);
44        mRemindIMGV=(ImageView) findViewById(R.id.imgv_traffic_remind);
45        mRemindTV=(TextView) findViewById(R.id.tv_traffic_remind);
46    }
47    private void initData(){
48        long totalflow=mSP.getLong("totalflow",0);
49        long usedflow=mSP.getLong("usedflow",0);
50        if(totalflow>0 & usedflow>=0){
51            float scale=usedflow/totalflow;
52            if(scale>0.9){
53                mRemindIMGV.setEnabled(false);
54                mRemindTV.setText("您的套餐流量即将用完!");
55            }else{
56                mRemindIMGV.setEnabled(true);
57                mRemindTV.setText("本月流量充足请放心使用");
58            }
59        }
60        mTotalTV.setText("本月流量: "+Formatter.formatFileSize(this,totalflow));
61        mUsedTV.setText("本月已用: "+Formatter.formatFileSize(this,usedflow));
62        dao=new TrafficDao(this);
63        Date date=new Date();
64        SimpleDateFormat sdf=new SimpleDateFormat("yyyy-MM-dd");
65        String dataString=sdf.format(date);
66        long moblieGPRS=dao.getMoblieGPRS(dataString);
67        if(moblieGPRS<0){
```

```java
68        moblieGPRS=0;
69    }
70    mToDayTV.setText("本日已用: "+Formatter.formatFileSize(this, moblieGPRS));
71  }
72  private void registReceiver(){
73      receiver=new CorrectFlowReceiver();
74      IntentFilter filter=new IntentFilter();
75      filter.addAction("android.provider.Telephony.SMS_RECEIVED");
76      registerReceiver(receiver,filter);
77  }
78  @Override
79  public void onClick(View v){
80      switch(v.getId()){
81      case R.id.imgv_leftbtn:
82          finish();
83          break;
84      case R.id.btn_correction_flow:
85          //首先判断是哪个运营商
86          int i=mSP.getInt("operator",0);
87          SmsManager smsManager=SmsManager.getDefault();
88          switch(i){
89          case 0:
90              //没有设置运营商
91              Toast.makeText(this,"您还没有设置运营商信息",0).show();
92              break;
93          case 1:
94              //中国移动
95              break;
96          case 2:
97              //中国联通,发送LLCX至10010
98              //获取系统默认的短信管理器
99              smsManager.sendTextMessage("10010", null, "LLCX", null, null);
100             break;
101         case 3:
102             //中国电信
103             break;
104         }
105     }
106 }
```

```java
107  class CorrectFlowReceiver extends BroadcastReceiver{
108   @Override
109   public void onReceive(Context context,Intent intent){
110     Object[] objs=(Object[]) intent.getExtras().get("pdus");
111     for(Object obj:objs){
112       SmsMessage smsMessage=SmsMessage.createFromPdu((byte[]) obj);
113       String body=smsMessage.getMessageBody();
114       String address=smsMessage.getOriginatingAddress();
115       //以下短信分割只针对联通3G用户
116       if(!address.equals("10010")){
117         return;
118       }
119       String[] split=body.split(", ");
120       //本月剩余流量
121       long left=0;
122       //本月已用流量
123       long used=0;
124       //本月超出流量
125       long beyond=0;
126       for(int i=0;i<split.length;i++){
127         if(split[i].contains("本月流量已使用")){
128           //套餐总量
129           String usedflow = split[i].substring(7,split[i].length());
130           used=getStringTofloat(usedflow);
131         } else if(split[i].contains("剩余流量")){
132           String leftflow = split[i].substring(4, split[i].length());
133           left=getStringTofloat(leftflow);
134         } else if(split[i].contains("套餐外流量")){
135           String beyondflow = split[i].substring(5, split[i].length());
136           beyond=getStringTofloat(beyondflow);
137         }
138       }
139       Editor edit=mSP.edit();
140       edit.putLong("totalflow",used+left);
141       edit.putLong("usedflow",used+beyond);
142       edit.commit();
143       mTotalTV.setText("本月流量: "+
144       Formatter.formatFileSize(context,(used+left)));
145       mUsedTV.setText("本月已用: "+
```

```
146            Formatter.formatFileSize(context,(used+beyond)));
147        }
148    }
149 }
150 /** 将字符串转化成Float类型数据 */
151 private long getStringTofloat(String str){
152    long flow=0;
153    if(!TextUtils.isEmpty(str)){
154        if(str.contains("KB")){
155            String[] split=str.split("KB");
156            float m=Float.parseFloat(split[0]);
157            flow=(long) (m*1024);
158        } else if(str.contains("MB")){
159            String[] split=str.split("MB");
160            float m=Float.parseFloat(split[0]);
161            flow=(long) (m*1024*1024);
162        } else if(str.contains("GB")){
163            String[] split=str.split("GB");
164            float m=Float.parseFloat(split[0]);
165            flow=(long) (m*1024*1024*1024);
166        }
167    }
168    return flow;
169 }
170 @Override
171 public void onDestroy(){
172    if(receiver!=null){
173        unregisterReceiver(receiver);
174        receiver=null;
175    }
176    super.onDestroy();
177 }
178 }
```

代码说明：

- 第20~23行代码获取SharedPreferences对象中存储的key为isset_operator的值，如果该值为false，则说明没有进入过运营商设置界面，并进入该界面。
- 第24~27行代码用于判断TrafficMonitoringService服务是否开启，如果没有开启，则启动服务。该服务主要功能为获取当前已使用流量，具体在8.4.3小节中讲解。
- 第47~71行的initData()方法获取流量使用的具体情况并将信息展示在界面中。第50~

59 行代码通过 SharedPreferences 对象获取到通过流量校验和在服务中获取到的流量数据，并计算出剩余流量的比例，通过比例进行判断，已用流量比例小于或者大于 90%，则显示不同的文字提示。第 60～61 行代码将获取到的月总流量和本月已用流量展示在界面中。第 62～70 行代码先获取当前时间，通过当前时间获取数据库中存储的当天的流量并展示在界面中。

- 第 72～77 行的 registReceiver()方法为注册自定义广播接收者，action 注册了接收短信的活动；
- 第 79～106 行的 onClick()方法用于监听校正流量按钮的点击事件，按下按钮后首先判断手机号码属于哪一个运营商，如果没有设置运营商信息，则提示进行设置。由于该模块只实现了联通的校对流量的方法，因此如果手机号码是联通，则使用系统的短信管理器向 10010 发送短信进行查询流量信息，移动和电信的实现方法大家可以模仿联通的实现方法进行编写。
- 第 107～149 行代码是自定义的广播接收者 CorrectFlowReceiver，当有短信到来时会执行 onReceive()方法。第 110～114 行代码获取短信的内容和电话号码；第 116～118 行代码判断如果不是联通官方服务电话 10010，则直接返回不再向下执行；第 119～138 行代码如果是联通官方服务电话，则对短信内容进行截取，分别获取短信内容中的本月已使用流量、剩余流量和套餐外流量，具体截取方法可以参考具体代码；第 139～146 行代码获取到流量信息后使用 SharedPreferences 进行存储，并显示在 TextView 上。

8.4.3 判断服务是否运行

在 8.4.2 小节中，使用了一个 SystemInfoUtils.isServiceRunning()方法，该方法用于判断服务是否处于运行状态（第 7 章中也使用过该方法），具体代码如【文件 8-11】所示。

【文件 8-11】SystemInfoUtils.java

```
1  public class SystemInfoUtils {
2      /**
3       * 判断一个服务是否处于运行状态
4       * @param context 上下文
5       * @return
6       */
7      public static boolean isServiceRunning(Context context,String className){
8          ActivityManager am=(ActivityManager) context.getSystemService(
9          Context.ACTIVITY_SERVICE);
10         List<RunningServiceInfo> infos=am.getRunningServices(200);
11         for(RunningServiceInfo info:infos){
12             String serviceClassName=info.service.getClassName();
13             if(className.equals(serviceClassName)){
14                 return true;
15             }
16         }
```

```
17        return false;
18    }
19 }
```

8.4.4 获取流量的服务

流量监控界面的逻辑代码中有一个判断，如果 TrafficMonitoringService 服务没有开启，则开启服务。该服务的主要作用是获取应用程序的实时流量信息，具体代码如【文件 8-12】所示。

【文件 8-12】TrafficMonitoringService.java

```java
1  public class TrafficMonitoringService extends Service {
2      private long mOldRxBytes;
3      private long mOldTxBytes;
4      private TrafficDao dao;
5      private SharedPreferences mSp;
6      private long usedFlow;
7      boolean flag=true;
8      @Override
9      public IBinder onBind(Intent intent){
10         return null;
11     }
12     @Override
13     public void onCreate(){
14         super.onCreate();
15         mOldRxBytes=TrafficStats.getMobileRxBytes();
16         mOldTxBytes=TrafficStats.getMobileTxBytes();
17         dao=new TrafficDao(this);
18         mSp=getSharedPreferences("config",MODE_PRIVATE);
19         mThread.start();
20     }
21     private Thread mThread=new Thread() {
22         public void run(){
23             while(flag){
24                 try{
25                     Thread.sleep(2000*60);
26                 }catch(InterruptedException e){
27                     e.printStackTrace();
28                 }
29                 updateTodayGPRS();
30             }
31         }
```

```java
32     private void updateTodayGPRS(){
33         //获取已经使用了的流量
34         usedFlow=mSp.getLong("usedflow",0);
35         Date date=new Date();
36         Calendar calendar=Calendar.getInstance();        //得到日历
37         calendar.setTime(date);                           //把当前时间赋给日历
38         if(calendar.DAY_OF_MONTH==1 & calendar.HOUR_OF_DAY==0
39             & calendar.MINUTE<1 & calendar.SECOND<30){
40             usedFlow=0;
41         }
42         SimpleDateFormat sdf=new SimpleDateFormat("yyyy-MM-dd");
43         String dataString=sdf.format(date);
44         long moblieGPRS=dao.getMoblieGPRS(dataString);
45         long mobileRxBytes=TrafficStats.getMobileRxBytes();
46         long mobileTxBytes=TrafficStats.getMobileTxBytes();
47         //新产生的流量
48         long newGprs=(mobileRxBytes+mobileTxBytes)-mOldRxBytes- mOldTxBytes;
49         mOldRxBytes=mobileRxBytes;
50         mOldTxBytes=mobileTxBytes;
51         if(newGprs<0){
52             //网络切换过
53             newGprs=mobileRxBytes+mobileTxBytes;
54         }
55         if(moblieGPRS==-1){
56             dao.insertTodayGPRS(newGprs);
57         }else{
58             if(moblieGPRS<0){
59                 moblieGPRS=0;
60             }
61             dao.updateTodayGPRS(moblieGPRS+newGprs);
62         }
63         usedFlow=usedFlow+newGprs;
64         Editor edit=mSp.edit();
65         edit.putLong("usedflow",usedFlow);
66         edit.commit();
67     };
68     };
69     @Override
70     public void onDestroy(){
```

```
71          if(mThread!=null & !mThread.interrupted()){
72              flag=false;
73              mThread.interrupt();
74              mThread=null;
75          }
76          super.onDestroy();
77      }
78 }
```

代码说明：

- 第 13～20 行 onCreate()方法在该服务刚启动时统计一次从本次开机到现在所使用的总流量，并运行子线程。
- 第 21～68 行创建了一个子线程，在线程中获取实时使用的流量信息。其中 22～31 行代码使线程睡眠 120 秒，该方法是一个子线程，并使用 while 循环使线程不断运行，为了电量及 CPU 使用的优化，在这里使线程睡眠 120 秒，不需要每一秒都在运行；第 32～68 行 updateTodayGPRS()方法向数据库中插入和更新最新的流量，并将总共使用的流量数据存储到 SharedPreferences 对象中。
- 第 34 行代码获取已经存储的所使用的总流量数据；
- 第 35～41 行代码，如果当前时间为每个月第一天的开始，则将所获取的当月使用流量信息清零。
- 第 42～44 行代码根据获得的当前时间查询数据库，获取今天的流量信息。
- 第 45～50 行代码先获取本次开机到现在的发送流量 mOldRxBytes 和接收流量 mOldTxBytes，两者相加再减去上次储存的流量就是本次服务到现在为止新产生的流量，再把这次获取的总流量赋值给服务刚开始运行时获取的总流量，并在每次循环时都将最新的流量数据赋值给变量 mOldRxBytes 和 mOldTxBytes。
- 第 51～54 行代码如果新产生的流量为 0 说明之前没有存储过流量，就将从本次开机到现在所产生的所有发送和接收的流量相加赋值给新产生的流量。
- 第 55～62 行代码判断获得的今天的流量为–1 时代表之前没有存储过流量，则将新产生的流量直接插入到数据库中，否则将今天的总流量加上新产生的流量更新到数据库。
- 第 63～66 行代码在服务刚创建时获取从本次开机到现在的总流量。
- 第 70～77 行代码当服务销毁时关闭线程并将线程销毁。

需要注意的是，通过服务可以获取实时流量数据，但如果用户在界面上点击"流量校正"按钮之后，在界面上显示的流量是以运营商短信为准的，并把通过短信获取的流量数据更新到 SharedPreferences 对象中。

接下来在 AndroidManifest.xml 文件中配置该服务的信息，具体代码如下所示：

```
<!-- 监控流量的服务 -->
<service
    android:name="cn.itcast.mobliesafe.chapter08.service.
    TrafficMonitoringService" android:persistent="true">
</service>
```

8.4.5 开机广播

获取流量的服务已经创建好，为了保证流量实时更新，在流量统计模块打开时就开启了获取流量服务。另外，在手机刚开机时也需要开启该服务。由于手机开机时系统会发送一条广播消息，因此可以利用广播接收者来开启服务，具体代码如【文件 8-13】所示。

【文件 8-13】BootCompleteReciever.java

```
1  public class BootCompleteReciever extends BroadcastReceiver{
2      @Override
3      public void onReceive(Context context,Intent intent){
4          //开机广播，判断流量监控服务是否开启，如果没开启则开启
5          if(!SystemInfoUtils.isServiceRunning(context, "cn.itcast.
6          mobliesafe.chapter08.service.TrafficMonitoringService")){
7              //开启服务
8              context.startService(new Intent(context,TrafficMonitoringService.class));
9          }
10     }
11 }
```

接下来在 AndroidManifest.xml 文件中配置开机启动的广播接收者，具体代码如下：

```
<receiver android:name="cn.itcast.mobliesafe.chapter08.reciever.
BootCompleteReciever" >
    <intent-filter>
        <action android:name="android.intent.action.BOOT_COMPLETED"/>
    </intent-filter>
</receiver>
```

本 章 小 结

本章主要是针对流量统计模块进行讲解，首先讲解了如何设置运营商信息，然后创建一个存储流量信息的数据库，最后讲解了获取并展示流量监控的逻辑。该模块逻辑较为复杂，尤其是计算流量时，编程者要慢慢体会并完全掌握。

【面试精选】
1. 请问使用 SQLite 数据库时有几种优化方式？
2. 请问 Activity 有几种启动模式，以及每种模式的特点？
扫描右方二维码，查看面试题答案！

第 9 章

高级工具模块

学习目标

- 了解高级工具模块功能；
- 掌握自定义控件的使用；
- 掌握第三方数据库的使用；
- 掌握如何进行短信备份与还原。

高级工具模块包含多种复杂而常用的功能，其中包括号码归属地查询、短信备份和还原及程序锁功能。通过该模块的学习，编程者可以掌握自定义控件、第三方数据库、服务以及内容观察者的使用。本章将针对高级工具模块进行详细讲解。

9.1 模块概述

9.1.1 功能介绍

高级工具模块包含四个独立功能，分别为号码归属地查询、短信备份和短信还原、程序锁。需要注意的是，虽然该模块讲解了四个功能，但是短信备份和短信还原这两个功能需要结合使用才是一个完整功能。接下来针对各个功能进行介绍。

1. 号码归属地查询

号码归属地查询功能是通过数据库查询出手机号码所在的城市以及运营商，具体如图 9-1 所示。

图 9-1　号码归属地查询界面

2. 短信备份和还原

短信备份功能就是将手机中的短信以 XML 格式保存到本地，这样即使不小心将短信删除，也可以通过短信还原功能从本地还原到系统短信数据库中，具体如图 9-2 所示。

图 9-2　短信备份和还原界面

3. 程序锁

每个人手机上都有一些隐私信息不想被别人看到，如短信、照片等，为此可以使用程序锁功能将其保护起来。程序锁可以将某一个应用上锁，当进入该程序时需要输入手机防盗密码，具体如图 9-3 所示。

图 9-3　程序锁界面

需要注意的是，在程序锁界面中使用 Fragment 控件分别展示未加锁应用界面和已加锁应用界面。将一个应用加锁后，会将该应用存入到数据库中。当在设置中心打开程序锁服务时，运行在后台的服务监测当前打开的应用，如果打开的应用在数据库中，就说明程序是加锁程序，会弹出输入密码界面，密码输入正确后才能打开应用（密码为手机防盗密码）。

9.1.2 代码结构

高级工具模块功能较为复杂，代码量大，为了让大家对该模块代码有个整体的认识，接下来通过一个图例来展示高级工具模块的代码结构，如图 9-4 所示。

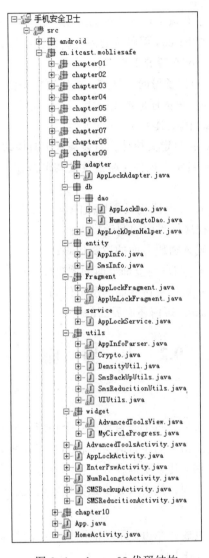

图 9-4　chapter09 代码结构

下面按照结构顺序依次介绍 chapter09 包中的文件，具体如下：
* AppLockAdapter.java：未加锁和已加锁界面 ListView 控件的数据适配器；
* AppLockDao.java：程序锁数据库的操作逻辑类；
* NumBelongtoDao.java：查询号码归属地的数据库逻辑类；
* AppLockOpenHelper.java：创建程序锁数据库；
* AppInfo.java：应用程序的实体类，包含 packageName、icon、appName、apkPath、isLock 字段；
* SmsInfo.java：短信的实体类，包含 body、date、type、address 字段；

- AppLockFragment.java：已加锁应用的界面逻辑；
- AppUnLockFragment.java：未加锁应用的界面逻辑；
- AppLockService.java：程序锁服务，判断应用是否已加锁，如果应用是加锁状态，打开应用时则显示输入密码框。
- AppInfoParser.java：用于获取手机中所有的应用程序的工具类；
- Crypto.java：加密和解密 base64 编码后的内容，用作程序锁密码；
- DensityUtil.java：dp 与 px 互转的工具类；
- SmsBackUpUtils.java：短信备份的工具类，提供短信备份 API；
- SmsReducitionUtils.java：短信还原的工具类，提供短信还原 API；
- UIUtils.java：封装 Toast 的工具类；
- AdvancedToolsView.java：自定义控件，封装高级工具模块的主页面；
- MyCircleProgress.java：自定义控件（带进度的圆形按钮），用于显示短信备份与还原时的进度；
- AdvancedToolsActivity.java：高级工具主界面的逻辑代码；
- AppLockActivity.java：程序锁界面的逻辑代码；
- EnterPswActivity.java：输入密码界面的逻辑代码；
- NumBelongtoActivity.java：归属地查询界面的逻辑代码；
- SMSBackupActivity.java：短信备份界面的逻辑代码；
- SMSReducitionActivity.java：短信还原界面的逻辑代码。

9.2 主 界 面

9.2.1 自定义组合控件

高级工具的主界面有四个条目，每个条目代表一个功能，分别是号码归属地查询、短信备份、短信还原和程序锁，当点击不同条目时会跳转到相应界面。由于这四个条目的布局都是一样的，如果使用普通布局相对比较麻烦，并且代码不宜编写和维护，因此该界面中的条目布局可以通过自定义控件来实现，接下来分步骤实现自定义控件。

1. 创建条目布局

每个条目布局都由三部分组成，左边显示功能图标，中间显示功能名称，右边显示一个箭头符号，具体代码如【文件 9-1】所示。

【文件 9-1】ui_advancedtools_view.xml

```xml
<?xml version="1.0" encoding="utf-8"?>
<RelativeLayout xmlns:android="http://schemas.android.com/apk/res/android"
    android:layout_width="match_parent"
    android:layout_height="55dp" >
    <ImageView
        android:id="@+id/imgv_left"
```

```xml
        android:layout_width="wrap_content"
        android:layout_height="wrap_content"
        android:layout_alignParentLeft="true"
        android:layout_centerVertical="true"
        android:layout_marginLeft="10dp"/>
    <ImageView
        android:id="@+id/imageView1"
        android:layout_width="13dp"
        android:layout_height="20dp"
        android:layout_alignParentRight="true"
        android:layout_centerVertical="true"
        android:layout_marginRight="20dp"
        android:background="@drawable/right_arrows_red"/>
    <TextView
        android:id="@+id/tv_decription"
        style="@style/textview16sp"
        android:layout_centerVertical="true"
        android:layout_marginLeft="30dp"
        android:layout_toRightOf="@+id/imgv_left"/>
    <View
        android:layout_width="match_parent"
        android:layout_height="1.0px"
        android:background="@color/black30"
        android:layout_alignParentBottom="true"/>
</RelativeLayout>
```

在上述代码中，TextView 控件使用了自定义样式 textview16sp，该样式在 styles.xml 中已经定义过了，可以直接引用（详细代码参见第 2 章）。View 控件使用了自定义颜色属性 black30，该颜色为黑色，具体代码如【文件 9-2】所示。

【文件 9-2】res/values/color.xml

```xml
<color name="black30">#30000000</color>
```

2. 自定义属性

在 res/values 目录下建立一个 attrs.xml 的文件（如果存在，则直接使用即可），该文件中添加控件的自定义属性，这样 R 文件中就可以生成对应的资源 id，就能在自定义控件中引用，具体代码如【文件 9-3】所示。

【文件 9-3】res/values/attrs.xml

```xml
<?xml version="1.0" encoding="utf-8"?>
<resources>
```

```
    <declare-styleable name="AdvancedToolsView">
        <attr name="desc" format="string" />
        <attr name="android:src"/>
    </declare-styleable>
</resources>
```

上述代码中，定义了 desc 和 android:src 两个属性，这两个属性在 AdvancedToolsView 类的第 16～19 行进行了调用。其中 desc 中的 format 字段表示属性的数据类型，可以为 String，也可以指向 R 文件中的索引等。

> **多学一招**：attrs.xml 文件的作用
>
> 在自定义一个控件时，如果需要定义一些新的属性，就会用到该 attrs.xml 文件，该文件中定义的是类的属性，这些属性是为了能在 XML 文件中被引用到，换句话说就是指定类中变量的值，这些属性会在类的构造方法中用到。看过一两个源码就会明白，构造方法中的 TypedArray 其实就是属性的数组，数组的成员会被赋给类中的成员，完成从 XML 的初始化。
>
> 在自定义控件的类中，通过构造方法 Class(Context context, AttributeSet attrs)和 Class（Context context, AttributeSet attrs, int defStyle)引入 attrs.xml 加载自定义属性，完成自定义控件的创建。

3. 创建自定义控件类

自定义控件类需要继承系统布局或者控件，并使用带 AttributeSet 参数的类的构造方法，在构造方法中将自定义控件类中变量与 attrs.xml 中的属性连接起来，具体代码如【文件 9-4】所示。

【文件 9-4】AdvancedToolsView.java

```
1  public class AdvancedToolsView extends RelativeLayout{
2      private TextView mDesriptionTV;
3      private String desc="";
4      private Drawable drawable;
5      private ImageView mLeftImgv;
6      public AdvancedToolsView(Context context) {
7          super(context);
8          init(context);
9      }
10     public AdvancedToolsView(Context context,AttributeSet attrs){
11         super(context, attrs);
12         //获取到属性对象
13         TypedArray mTypedArray=context.obtainStyledAttributes(attrs,
14         R.styleable.AdvancedToolsView);
15         //获取到 desc 属性，与 attrs.xml 中定义的属性绑定
```

```
16      desc=mTypedArray.getString(R.styleable.AdvancedToolsView_desc);
17      //获取到android:src属性,与attrs.xml中定义的属性绑定
18      drawable=mTypedArray.getDrawable(R.styleable.
19      AdvancedToolsView_android_src);
20      //回收属性对象
21      mTypedArray.recycle();
22      init(context);
23   }
24   public AdvancedToolsView(Context context,AttributeSet attrs,int defStyle){
25      super(context,attrs,defStyle);
26      init(context);
27   }
28   /**
29    * 控件初始化
30    * @param context
31    */
32   private void init(Context context){
33      //将资源转化成view对象显示在自己身上
34      View view = View.inflate(context, R.layout.ui_advancedtools_view, null);
35      this.addView(view);
36      mDesriptionTV=(TextView) findViewById(R.id.tv_decription);
37      mLeftImgv=(ImageView) findViewById(R.id.imgv_left);
38      mDesriptionTV.setText(desc);
39      if(drawable!=null)mLeftImgv.setImageDrawable(drawable);
40   }
41 }
```

上述代码中,首先创建一个 AdvancedToolsView 类继承 RelativeLayout,并实现该类的三个构造方法。其中第一个构造方法 AdvancedToolsView(Context context)是通过继承已有的控件来实现自定义控件,主要是当要实现的控件和已有控件在很多方面比较相似,通过对已有控件的扩展来满足要求时使用。第二个构造方法 AdvancedToolsView(Context context, AttributeSet attrs)通过继承一个布局文件实现自定义控件,一般来说做组合控件时可以使用这种方式实现(在 XML 创建但是没有指定 style 的时候调用)。第三个构造方法 AdvancedToolsView(Context context,AttributeSet attrs, int defStyle)通过继承 View 来实现自定义控件,使用 GDI 绘制出组件界面,一般无法通过上述两种方式来实现时用该方式,当有指定的 style 时调用该方法。由于此处是通过布局文件来实现自定义控件的,因此需要使用第二个构造方法。

4. 布局中使用自定义控件

在布局中引用已经创建好的自定义控件,具体效果图如图 9-5 所示。

图 9-5 所示高级工具主界面所对应的布局文件如【文件 9-5】所示。

【文件 9-5】activity_advancetools.xml

图 9-5 高级工具主界面

```xml
<?xml version="1.0" encoding="utf-8"?>
<LinearLayout xmlns:android="http://schemas.android.com/apk/res/android"
    xmlns:custom_android="http://schemas.android.com/apk/res/cn.itcast.mobliesafe"
    android:layout_width="match_parent"
    android:layout_height="match_parent"
    android:orientation="vertical">
    <include layout="@layout/titlebar"/>
    <cn.itcast.mobliesafe.chapter09.widget.AdvancedToolsView
        android:id="@+id/advanceview_numbelongs"
        android:layout_width="match_parent"
        android:layout_height="55dp"
        custom_android:desc="号码归属地查询"
        android:gravity="center_vertical"
        android:src="@drawable/numofbelongto">
    </cn.itcast.mobliesafe.chapter09.widget.AdvancedToolsView>
    <cn.itcast.mobliesafe.chapter09.widget.AdvancedToolsView
        android:id="@+id/advanceview_smsbackup"
        android:layout_width="match_parent"
        android:layout_height="55dp"
        custom_android:desc="短信备份"
        android:gravity="center_vertical"
        android:layout_marginTop="3dp"
        android:src="@drawable/sms_backup">
    </cn.itcast.mobliesafe.chapter09.widget.AdvancedToolsView>
    <cn.itcast.mobliesafe.chapter09.widget.AdvancedToolsView
        android:id="@+id/advanceview_smsreducition"
        android:layout_width="match_parent"
        android:layout_height="55dp"
        android:gravity="center_vertical"
        android:layout_marginTop="3dp"
        custom_android:desc="短信还原"
        android:src="@drawable/sms_reduction">
    </cn.itcast.mobliesafe.chapter09.widget.AdvancedToolsView>
    <cn.itcast.mobliesafe.chapter09.widget.AdvancedToolsView
```

```
            android:id="@+id/advanceview_applock"
            android:layout_width="match_parent"
            android:layout_height="55dp"
            android:gravity="center_vertical"
            android:layout_marginTop="3dp"
            custom_android:desc="程序锁"
            android:src="@drawable/programoflock">
    </cn.itcast.mobliesafe.chapter09.widget.AdvancedToolsView>
</LinearLayout>
```

在使用自定义控件时,每个控件的开始节点和结束节点都为自定义控件类的全路径,其中 custom_android:desc 属性和 android:src 属性是自定义的,分别用于显示描述文字以及功能图标。这样无论再添加多少个条目也只是多添加几个自定义控件,这样更利于程序编写和维护。

需要注意的是,在使用自定义控件时必须使用 XML 命名空间 xmlns:custom_android="http://schemas.android.com/apk/res/cn.itcast.mobliesafe"将自定义控件引入到布局。

至此,自定义控件已经完成。要使用自定义控件,需要在具体类中创建该自定义控件类对象,然后对控件中属性进行操作。

9.2.2 主界面逻辑

主界面逻辑较简单,主要是设置每一个条目的点击事件,跳转到相应界面即可。在主界面中需要创建 9.2.1 小节中创建的自定义控件对象(自定义控件的第五步操作),具体代码如【文件 9-6】所示。

【文件 9-6】AdvancedToolsActivity.java

```
1  public class AdvancedToolsActivity extends Activity implements OnClickListener{
2      @Override
3      protected void onCreate(Bundle savedInstanceState){
4          super.onCreate(savedInstanceState);
5          requestWindowFeature(Window.FEATURE_NO_TITLE);
6          setContentView(R.layout.activity_advancetools);
7          initView();
8      }
9      /**初始化控件*/
10     private void initView(){
11         findViewById(R.id.rl_titlebar).setBackgroundColor(
12             getResources().getColor(R.color.bright_red));
13         ImageView mLeftImgv=(ImageView) findViewById(R.id.imgv_leftbtn);
14         ((TextView) findViewById(R.id.tv_title)).setText("高级工具");
15         mLeftImgv.setOnClickListener(this);
```

```
16        mLeftImgv.setImageResource(R.drawable.back);
17        findViewById(R.id.advanceview_applock).setOnClickListener(this);
18        findViewById(R.id.advanceview_numbelongs).setOnClickListener(this);
19        findViewById(R.id.advanceview_smsbackup).setOnClickListener(this);
20        findViewById(R.id.advanceview_smsreducition).setOnClickListener(this);
21    }
22    @Override
23    public void onClick(View v){
24        switch(v.getId()){
25        case R.id.imgv_leftbtn:
26            finish();
27            break;
28        case R.id.advanceview_applock:
29            //进入程序锁页面
30            startActivity(AppLockActivity.class);
31            break;
32        case R.id.advanceview_numbelongs:
33            //进入归属地查询页面
34            startActivity(NumBelongtoActivity.class);
35            break;
36        case R.id.advanceview_smsbackup:
37            //进入短信备份页面
38            startActivity(SMSBackupActivity.class);
39            break;
40        case R.id.advanceview_smsreducition:
41            //进入短信还原页面
42            startActivity(SMSReducitionActivity.class);
43            break;
44        }
45    }
46    /**
47     * 开启新的activity不关闭自己
48     * @param cls 新的activity的字节码
49     */
50    public void startActivity(Class<?> cls){
51        Intent intent=new Intent(this,cls);
52        startActivity(intent);
53    }
54 }
```

代码说明：

- 第 10～21 行的 initView()方法，用于初始化布局中的自定义控件，由于不需要直接对控件中的数据进行操作，因此不必要 new 出一个对象，直接对该控件添加点击监听即可。
- 第 23～45 行的 onClick()方法用于响应主界面 4 个条目的点击事件。
- 第 50～53 行的 startActivity()方法用于开启新的 Activity，单独将这段代码摘出，是因为四个条目所做的操作相同，因此可以直接复用该方法，以减少代码重复。

9.3 号码归属地查询

在查询号码归属地时，主要用到了一个第三方数据库，该数据库存储了大量的归属地信息，查询号码归属地功能实际上就是通过这个数据库实现的。本节将针对号码归属地查询进行详细讲解。

9.3.1 号码归属地查询界面

号码归属地查询界面非常简单，主要由一个输入框、一个查询按钮以及显示归属地信息的 TextView 组成，默认情况下显示号码归属地的 TextView 不显示，只有当输入框中输入查询号码，点击"查询"按钮时，才会显示。

图 9-6 所示号码归属地对应的布局文件如【文件 9-7】所示。

【文件 9-7】activity_numbelongto.xml

图 9-6　号码归属地查询界面

```
<?xml version="1.0" encoding="utf-8"?>
<LinearLayout
    xmlns:android="http://schemas.android.com/apk/res/android"
    android:layout_width="match_parent"
    android:layout_height="match_parent"
    android:orientation="vertical">
    <include layout="@layout/titlebar"/>
    <EditText
        android:id="@+id/et_num_numbelongto"
        android:layout_marginTop="20dp"
        android:layout_marginLeft="10dp"
        android:layout_marginRight="10dp"
        android:layout_width="match_parent"
        android:layout_height="wrap_content"
        android:inputType="phone"
        android:drawableLeft="@drawable/numbelongtoicon"
        android:background="@drawable/numbelongto_et_bg"
```

```xml
            android:drawablePadding="3dp"
            android:hint="请输入要查询的号码"/>
    <Button
        android:id="@+id/btn_searchnumbelongto"
        android:layout_marginTop="20dp"
        android:layout_marginLeft="10dp"
        android:layout_marginRight="10dp"
        android:layout_width="match_parent"
        android:layout_height="wrap_content"
        android:background="@drawable/btn_numaddressbg_selector"/>
    <TextView
        android:id="@+id/tv_searchresult"
        style="@style/textview16sp"
        android:layout_marginLeft="30dp"
        android:layout_marginTop="15dp"
        android:textColor="@color/deep_gray"/>
</LinearLayout>
```

上述布局较简单,不做过多描述,其中 Button 按钮使用了一个背景选择器,用于控制按钮的按下与弹起时的背景,具体代码如【文件 9-8】所示。

【文件 9-8】res/drawable/btn_numaddressbg_selector.xml

```xml
<?xml version="1.0" encoding="utf-8"?>
<selector xmlns:android="http://schemas.android.com/apk/res/android" >
    <item android:state_pressed="true"
        android:drawable="@drawable/btn_search_numbelongto_p"/>
    <item android:drawable="@drawable/btn_search_numbelongto_n"/>
</selector>
```

在上述代码中,当按钮按下时显示 btn_search_numbelongto_p.png(红色)图片,当按钮弹起时显示 btn_search_numbelongto_n.png(灰色)图片。

9.3.2 数据库展示

在开发号码归属地查询功能之前,需要先创建一个 address.db 数据库,用于存储内地所有省市的号码信息,这样通过查询数据库就可以查询出号码归属地。由于这个数据库所涉及的数据过多,创建起来很麻烦,因此,直接使用第三方的数据库 address.db。

address.db 数据库中有两个表 data1 和 data2,data1 表有三个字段 RecNo、id 和 outkey,其中 RecNo 是 SQLite 自带的自增字段,id 为主键用于存储电话号码的前 7 位,outkey 相当于外键,用于与 data2 表中的 id 关联。具体如图 9-7 所示。

data2 表中有四个字段 RecNo、id、location、area,其中 id 为自动增长的主键,location 用于存储号码归属地,area 用于存储座机的区号(去掉 0),具体如图 9-8 所示。

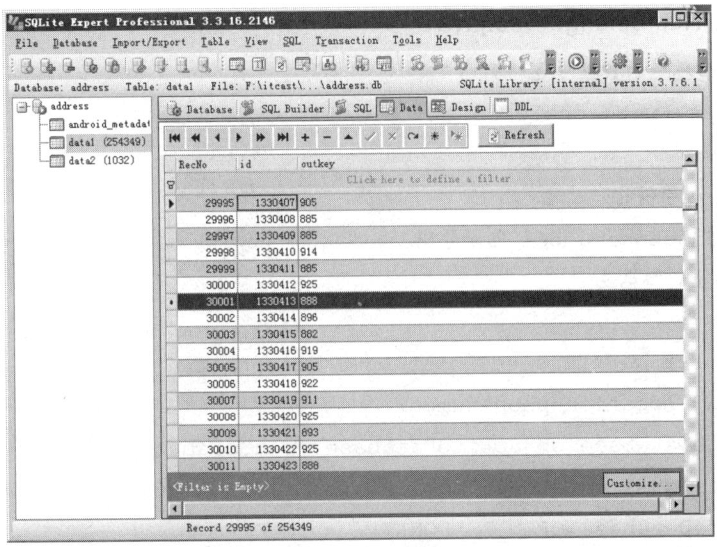

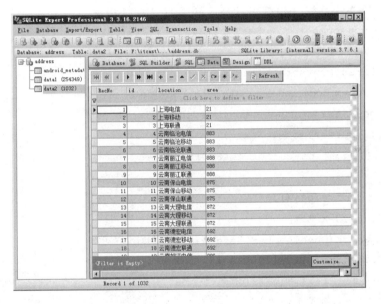

图 9-7　address.db-data1

图 9-8　address.db-data2

在查询号码归属地时，首先需要根据传入的电话号码截取前 7 位找到 data1 表中相对应的 outkey 值，然后将得到的 outkey 值作为 data2 表中的 id 查询相对应的 location 值，该值就是要得到的电话号码归属地。如果要查询座机号码就只需要使用 data2 表中的 area 字段，根据该字段查询对应的 location 值即可。

9.3.3　数据库操作

在进行号码归属地查询时，主要是通过电话号码与数据库中的数据进行比对实现的，因此，首先需要创建一个数据库操作类，在该类中通过 SQL 语句查询出号码的归属地，具体代码如【文件 9-9】所示。

【文件 9-9】NumBelongtoDao.java

```java
1  /**查询号码归属地的数据库逻辑类*/
2  public class NumBelongtoDao{
3      /**
4       * 返回电话号码的归属地
5       * @param phonenumber 电话号码
6       * @return 归属地
7       */
8      public static String getLocation(String phonenumber){
9          String location=phonenumber;
10         SQLiteDatabase db=SQLiteDatabase.openDatabase(
11         "/data/data/cn.itcast.mobliesafe/files/address.db",null,
12         SQLiteDatabase.OPEN_READONLY);
13         //通过正则表达式匹配号段，13X  14X  15X  17X  18X,
14         //130 131 132 133 134 135 136 137 137 139
15         if(phonenumber.matches("^1[34578]\\d{9}$")){
16             //手机号码的查询
17             Cursor cursor=db.rawQuery(
18             "select location from data2 where id=(select outkey from data1 where id=?)",
19             new String[] {phonenumber.substring(0,7)});
20             if(cursor.moveToNext()){
21                 location=cursor.getString(0);
22             }
23             cursor.close();
24         }else{//其他电话
25             switch(phonenumber.length()){//判断电话号码的长度
26             case 3: //110 120 119 121 999
27                 if("110".equals(phonenumber)){
28                     location="匪警";
29                 }elseif("120".equals(phonenumber)){
30                     location="急救";
31                 }else{
32                     location="报警号码";
33                 }
34                 break;
35             case 4:
36                 location="模拟器";
37                 break;
38             case 5:
```

```java
39              location="客服电话";
40              break;
41          case 7:
42              location="本地电话";
43              break;
44          case 8:
45              location="本地电话";
46              break;
47          default:
48              if(location.length()>=9 && location.startsWith("0")){
49                  String address=null;
50                  //select location from data2 where area='10'
51                  Cursor cursor=db.rawQuery("select location from data2 where
52                  area = ?", new String[]{location.substring (1, 3)});
53                  if(cursor.moveToNext()){
54                      String str=cursor.getString(0);
55                      address=str.substring(0,str.length()-2);
56                  }
57                  cursor.close();
58                  cursor = db.rawQuery("select location from data2 where area = ?",
59                          new String[]{location.substring(1,4)});
60                  if(cursor.moveToNext()){
61                      String str=cursor.getString(0);
62                      address=str.substring(0,str.length()-2);
63                  }
64                  cursor.close();
65                  if(!TextUtils.isEmpty(address)){
66                      location=address;
67                  }
68              }
69              break;
70          }
71      }
72      db.close();
73      return location;
74  }
75 }
```

代码说明：
- 第 9～12 行代码用于打开数据库，该数据库需要先从 assets 目录复制到工程目录的 files 目录下。
- 第 15～23 行代码使用正则表达式判断接收的号码是否为手机号码，如果是，则返回手机号码的归属地。
- 第 25～46 行代码判断电话号码的长度，不同长度的电话进行不同的处理，比如三位数的电话是报警电话，四位数的是模拟器，五位数是客服电话，七位和八位的是本地电话。
- 第 48～68 行代码判断座机号码，如果电话号码超过了 9 位并且是以 0 开头，则截取号码的第二、三位或者截取号码的第二、三、四位和数据库中 data2 表中 area 字段进行比对，查询出相同的 area 值。需要注意的是，同一个地区的运营商 area 值都是相同的，比如 010 开头的代表北京电信、北京联通和北京移动，这三个数据在数据库中 data2 表中 area 值都是"10"，无法精确判断出是三个运营商中的哪一种，因此得到结果之后再次进行截取，只保留结果的前两位"北京"。

9.3.4 号码归属地查询逻辑

号码归属地主逻辑代码比较简单，首先需要将 address.db 放到 assets 目录下，通过代码将数据库复制到工程目录中，然后在界面中调用数据库操作类中的 getLocation() 方法，即可查询出归属地，具体代码如【文件 9-10】所示。

【文件 9-10】NumBelongtoActivity.java

```
1  /**归属地查询*/
2  public class NumBelongtoActivity extends Activity implements OnClickListener{
3      private EditText mNumET;
4      private TextView mResultTV;
5      private String dbName="address.db";
6      private Handler mHandler=new Handler(){
7          public void handleMessage(android.os.Message msg){
8          };
9      };
10     @Override
11     protected void onCreate(Bundle savedInstanceState){
12         super.onCreate(savedInstanceState);
13         requestWindowFeature(Window.FEATURE_NO_TITLE);
14         setContentView(R.layout.activity_numbelongto);
15         initView();
16         copyDB(dbName);
17     }
18     /**初始化控件*/
19     private void initView(){
20         findViewById(R.id.rl_titlebar).setBackgroundColor(
```

```
21          getResources().getColor(R.color.bright_red));
22      ImageView mLeftImgv=(ImageView) findViewById(R.id.imgv_leftbtn);
23      ((TextView) findViewById(R.id.tv_title)).setText("号码归属地查询");
24      mLeftImgv.setOnClickListener(this);
25      mLeftImgv.setImageResource(R.drawable.back);
26      findViewById(R.id.btn_searchnumbelongto).setOnClickListener(this);
27      mNumET=(EditText) findViewById(R.id.et_num_numbelongto);
28      mResultTV=(TextView) findViewById(R.id.tv_searchresult);
29      mNumET.addTextChangedListener(new TextWatcher(){
30          @Override
31          public void onTextChanged(CharSequence s,int start,int before,int count){
32          }
33          @Override
34          public void beforeTextChanged(CharSequence s,int start,int count,
35          int after){
36          }
37          @Override
38          public void afterTextChanged(Editable s){
39              //文本变化之后
40              String string=s.toString().trim();
41              if(string.length()==0){
42                  mResultTV.setText("");
43              }
44          }
45      });
46  }
47  @Override
48  public void onClick(View v){
49      switch(v.getId()){
50      case R.id.imgv_leftbtn:
51          finish();
52          break;
53      case R.id.btn_searchnumbelongto:
54          //判断edittext中的号码是否为空
55          //判断数据库是否存在
56          String phonenumber=mNumET.getText().toString().trim();
57          if(!TextUtils.isEmpty(phonenumber)){
58              File file=new File(getFilesDir(),dbName);
59              if(!file.exists()||file.length()<=0){
```

```
60              //数据库不存在,复制数据库
61              copyDB(dbName);
62          }
63          //查询数据库
64          String location=NumBelongtoDao.getLocation(phonenumber);
65          mResultTV.setText("归属地: "+location);
66      }else{
67          Toast.makeText(this,"请输入需要查询的号码",0).show();
68      }
69      break;
70   }
71 }
72 /**
73  * 复制资产目录下的数据库文件
74  * @param dbname   数据库文件的名称
75  */
76 private void copyDB(final String dbname){
77    new Thread(){
78       public void run(){
79          try{
80             File file=new File(getFilesDir(),dbname);
81             if(file.exists() && file.length()>0){
82                Log.i("NumBelongtoActivity","数据库已存在");
83                return ;
84             }
85             InputStream is=getAssets().open(dbname);
86             FileOutputStream fos=openFileOutput(dbname, MODE_PRIVATE);
87             byte[] buffer=new byte[1024];
88             int len=0;
89             while((len=is.read(buffer))!=-1){
90                fos.write(buffer,0,len);
91             }
92             is.close();
93             fos.close();
94          } catch(Exception e){
95             e.printStackTrace();
96          }
97       };
98    }.start();
```

```
 99     }
100 }
```

代码说明：

- 第 19~46 行的 initView() 方法用于初始化界面，其中第 29~44 行代码为 EditText 的内容监听器，在文本框输入完成之后判断如果为空，就把归属地信息清空。
- 第 48~71 行的 onClick() 方法用于响应界面中按钮的点击事件，如果文本框中没有内容则给予提示先输入电话号码，然后查询数据库是否存在，如果不存在则先进行数据库复制，然后遍历数据库，查出号码归属地。
- 第 76~99 行的 copyDB() 方法用于复制数据库，在界面初始化时中调用，该逻辑在之前章节已经介绍过，这里不再赘述。

9.4 短信备份

9.4.1 短信备份工具类

要将短信备份在本地（内存或 SD 卡）有多种方法，在这里使用 XML 的形式保存短信内容。Android 手机中的短信是在 data/data/com.android.provider.telephony 应用 database 目录下的 mmssms.db 数据库中，如图 9-9 所示。

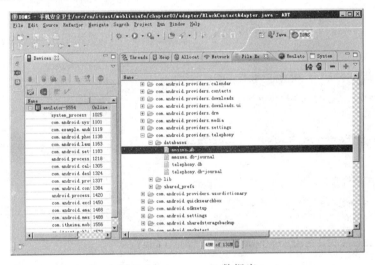

图 9-9 mmssms.db 数据库

通过 SQLiteExpert 工具打开 mmssms.db 数据库中的 sms 表，可以看到手机中的短信内容，具体如图 9-10 所示。

在图 9-10 的 sms 表中，只需要关心其中的几项数据即可，如 address 代表短信地址；type 代表短信类型：2 代表发送，1 代表接收；date 代表短信时间；body 代表短信内容。要将短信备份为 XML 文件，只需将这几项数据以节点形式写入 XML 文件即可。

在获取短信内容时，不直接操作数据库，而是使用 ContentResolver 得到短信数据库中的这几项数据，具体代码如【文件 9-11】所示。

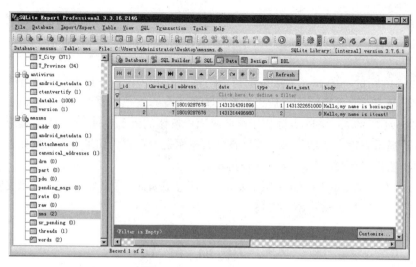

图 9-10 mmssms.db-sms

【文件 9-11】SmsBackUpUtils.java

```java
1  /**
2   * 短信的工具类，提供短信备份 API
3   * @author Administrator
4   */
5  public class SmsBackUpUtils{
6      /**
7       * 定义的一个接口，用作回调
8       * @author Administrator
9       */
10     public interface BackupStatusCallback{
11         /**
12          * 在短信备份之前调用的方法
13          * @param size 总的短信的个数
14          */
15         public void beforeSmsBackup(int size);
16         /**
17          * 当 sms 短信备份过程中调用的方法
18          * @param process 当前的进度
19          */
20         public void onSmsBackup(int process);
21     }
22     private boolean flag=true;
23     public void setFlag(boolean flag){
24         this.flag=flag;
```

```
25    }
26    /**
27     * 备份短信
28     * @param context 上下文
29     * @param callback 实现BackupStatusCallback接口的对象
30     * @return
31     * @throws FileNotFoundException
32     */
33    public boolean backUpSms(Context context,BackupStatusCallback callback)
34        throws FileNotFoundException,IllegalStateException,IOException{
35        XmlSerializer serializer=Xml.newSerializer();
36        File sdDir=Environment.getExternalStorageDirectory();
37        long freesize=sdDir.getFreeSpace();
38        if(Environment.getExternalStorageState().equals(
39           Environment.MEDIA_MOUNTED) && freesize>1024l*1024l){
40           File file=new File(Environment.getExternalStorageDirectory(),
41           "backup.xml");
42           FileOutputStream os=new FileOutputStream(file);
43           //初始化xml文件的序列化器
44           serializer.setOutput(os,"utf-8");
45           //写xml文件的头
46           serializer.startDocument("utf-8",true);
47           //写根节点
48           ContentResolver resolver=context.getContentResolver();
49           Uri uri=Uri.parse("content://sms/");
50           Cursor cursor=resolver.query(uri,new String[] {"address","body",
51           "type","date"},null,null,null);
52           //得到总的条目个数
53           int size=cursor.getCount();
54           //设置进度的总大小
55           callback.beforeSmsBackup(size);
56           serializer.startTag(null,"smss");
57           serializer.attribute(null,"size",String.valueOf(size));
58           int process=0;
59           while(cursor.moveToNext() & flag) {
60              serializer.startTag(null,"sms");
61              serializer.startTag(null,"body");
62              //可能会有乱码问题需要处理,如果出现乱码会导致备份失败
63              try {
```

```
64              String bodyencpyt=Crypto.encrypt("123",cursor.getString(1));
65              serializer.text(bodyencpyt);
66          } catch (Exception e1){
67              e1.printStackTrace();
68              serializer.text("短信读取失败");
69          }
70          serializer.endTag(null,"body");
71          serializer.startTag(null,"address");
72          serializer.text(cursor.getString(0));
73          serializer.endTag(null,"address");
74          serializer.startTag(null,"type");
75          serializer.text(cursor.getString(2));
76          serializer.endTag(null,"type");
77          serializer.startTag(null,"date");
78          serializer.text(cursor.getString(3));
79          serializer.endTag(null,"date");
80          serializer.endTag(null,"sms");
81          try{
82              Thread.sleep(600);
83          }catch(InterruptedException e){
84              e.printStackTrace();
85          }
86          //设置进度条对话框的进度
87          process++;
88          callback.onSmsBackup(process);
89      }
90      cursor.close();
91      serializer.endTag(null,"smss");
92      serializer.endDocument();
93      os.flush();
94      os.close();
95      return flag;
96  }else{
97      throw new IllegalStateException("sd卡不存在或者空间不足");
98  }
99  }
100 }
```

代码说明:

- 第5～21行代码定义一个回调接口BackupStatusCallback,接口中定义两个方法,分别在短信备份之前调用和短信备份过程中调用。其中 beforeSmsBackup(int size)方法在第 55 行得到短信总条数的时候进行设置,参数为接收短信的总条数,该方法用于主界面调用时进行是否备份短信的判断,如果总条数是 0 则不备份;onSmsBackup(int process)方法在第 88 行短信备份过程中进行设置,参数接收备份短信的进度,该方法用于主界面调用时的进度条的设置。

- 第 22~25 行的 setFlag()方法是在 SMSBackupActivity 界面中点击备份或取消按钮时调用,用来控制短信备份的开始和取消。
- 第 33~99 行的 backUpSms()方法用于备份短信,首先创建一个备份文件 backup.xml,然后获取到短信内容,并对其进行序列化,将短信内容存储在 backup.xml 文件中。其中第 48~51 行代码使用 ContentResolver 得到短信数据。第 64 行代码将 body 信息使用 base64 进行加密,防止其他人得到该文件造成短信内容泄露。

9.4.2 短信加密和解密

为防止导出在本地的短信内容被别人窃取和查看,可以使用 base64 对象将短信内容进行加密,当还原或者导入其他设备时同样需要使用相同密码将短信进行解密,以保证短信的私密性和安全性,具体代码如【文件 9-12】所示。

【文件 9-12】Crypto.java

```
1  public class Crypto{
2      /**
3       * 加密一个文本,返回base64编码后的内容
4       * @param seed   种子 密码
5       * @param plain  原文
6       * @return 密文
7       * @throws Exception
8       */
9      public static String encrypt(String seed,String plain) throws Exception {
10         byte[] rawKey=getRawKey(seed.getBytes());
11         byte[] encrypted=encrypt(rawKey,plain.getBytes());
12         return Base64.encodeToString(encrypted,Base64.DEFAULT);
13     }
14     /**
15      * 解密base64编码后的密文
16      * @param seed   种子 密码
17      * @param encrypted  密文
18      * @return 原文
19      * @throws Exception
20      */
21     public static String decrypt(String seed,String encrypted) throws Exception{
22         byte[] rawKey=getRawKey(seed.getBytes());
```

```java
23      byte[] enc=Base64.decode(encrypted.getBytes(),Base64.DEFAULT);
24      byte[] result=decrypt(rawKey, enc);
25      return new String(result);
26  }
27  private static byte[] getRawKey(byte[] seed) throws Exception{
28      KeyGenerator keygen=KeyGenerator.getInstance("AES");
29      SecureRandom random=SecureRandom.getInstance("SHA1PRNG");
30      random.setSeed(seed);
31      keygen.init(128,random);
32      SecretKey key=keygen.generateKey();
33      byte[] raw=key.getEncoded();
34      return raw;
35  }
36  private static byte[] encrypt(byte[] raw,byte[] plain) throws Exception{
37      SecretKeySpec keySpec=new SecretKeySpec(raw, "AES");
38      Cipher cipher=Cipher.getInstance("AES");
39      cipher.init(Cipher.ENCRYPT_MODE,keySpec);
40      byte[] encrypted=cipher.doFinal(plain);
41      return encrypted;
42  }
43  private static byte[] decrypt(byte[] raw,byte[] encrypted) throws Exception{
44      SecretKeySpec keySpec=new SecretKeySpec(raw,"AES");
45      Cipher cipher=Cipher.getInstance("AES");
46      cipher.init(Cipher.DECRYPT_MODE,keySpec);
47      byte[] decrypted=cipher.doFinal(encrypted);
48      return decrypted;
49  }
50}
```

代码说明：

- 第9~13行的encrypt(String seed, String plain)方法用于加密文本，该方法接收两个参数，第一个参数为密码，解密时必须使用和加密时一样的密码，第二个参数接收需要加密的原文。
- 第21~26行的decrypt(String seed, String encrypted)方法用于解密文本，该方法接收两个参数，第一个参数为密码，该密码必须和加密时的密码相同，第二个参数为已经加密过的密文。

9.4.3 短信备份界面

短信备份界面使用了自定义控件，界面中间有一个红色按钮，点击之后开始备份短信，并显示备份进度。由于要在自定义控件中使用自定义的属性，显示备份进度、颜色等，因此先要在res/values目录的attrs.xml文件中创建对应的变量，具体如【文件9-13】所示。

【文件 9-13】res/values/attrs.xml

```xml
<?xml version="1.0" encoding="utf-8"?>
<resources>
    <declare-styleable name="MyCircleProgress">
        <attr name="progress" format="integer"/>
        <attr name="max" format="integer"/>
        <attr name="circleColor" format="color"/>
        <attr name="progressColor" format="color"/>
        <attr name="android:background"/>
    </declare-styleable>
</resources>
```

在上述代码所示，MyCircleProgress 是自定义控件的根节点，它有五个属性，分别是设置进度的 progress，设置进度最大值的 max，设置圆形颜色的 circleColor，设置进度条颜色的 progressColor，设置背景的 android:background。

所有准备工作已完成，接下来自定义短信备份按钮控件，具体代码如【文件 9-14】所示。

【文件 9-14】MyCircleProgress.java

```java
1  /**自定义控件 带进度的按钮*/
2  public class MyCircleProgress extends Button{
3      private Paint paint;
4      /**进度*/
5      private int progress;
6      private int max;
7      /**进度为零时，颜色*/
8      private int mCircleColor;
9      private int mProgressColor;
10     private float roundWidth;
11     private int mProgressTextSize;
12     private Context context;
13     private float mDistanceOFbg;
14     public MyCircleProgress(Context context){
15         this(context,null);
16     }
17     public MyCircleProgress(Context context,AttributeSet attrs){
18         this(context,attrs,0);
19     }
20     public MyCircleProgress(Context context,AttributeSet attrs,int defStyle){
21         super(context,attrs,defStyle);
22         init(context,attrs);
```

```java
}
/**
 * 初始化控件
 * @param context
 */
private void init(Context context,AttributeSet attrs){
    paint=new Paint();
    this.context=context;
    TypedArray typedArray=context.obtainStyledAttributes(
    attrs,R.styleable.MyCircleProgress);
    progress=typedArray.getInteger(R.styleable.MyCircleProgress_progress,0);
    max=typedArray.getInteger(R.styleable.MyCircleProgress_max,100);
    mCircleColor=typedArray.getColor(R.styleable.MyCircleProgress_
    circleColor,Color.RED);
    mProgressColor=typedArray.getColor(
    R.styleable.MyCircleProgress_progressColor,Color.WHITE);
    roundWidth=DensityUtil.dip2px(context,5);
    mDistanceOFbg=DensityUtil.dip2px(context,5);
    mProgressTextSize=DensityUtil.dip2px(context,16);
    typedArray.recycle();
}
@Override
protected void onDraw(Canvas canvas){
    super.onDraw(canvas);
    //步骤1: 计算出圆心的位置
    int centerX=getWidth()/2;
    int centerY=getHeight()/2;
    int radius=(int) (centerX-mDistanceOFbg);
    //步骤2: 画出外层圆圈
    paint.setColor(mCircleColor);              //设置颜色
    paint.setAntiAlias(true);                  //给Paint加上锯齿
    paint.setStyle(Paint.Style.STROKE);        //设置填充样式,仅描边
    paint.setStrokeWidth(roundWidth);          //设置画笔宽度
    canvas.drawCircle(centerX,centerY,radius,paint); //画圆
    //步骤3: 画出外层进度条
    paint.setColor(mProgressColor);
    paint.setAntiAlias(true);
    paint.setStyle(Paint.Style.STROKE);
    paint.setStrokeWidth(roundWidth);
```

```java
        //用于定义的圆弧的形状和大小的界限
        RectF oval=new RectF(centerX-radius,centerY-radius,centerX+radius,
        centerY+radius);
        paint.setStyle(Paint.Style.STROKE);
        //旋转后,图片的抗锯齿
        canvas.setDrawFilter(new PaintFlagsDrawFilter(0,
        Paint.ANTI_ALIAS_FLAG|Paint.FILTER_BITMAP_FLAG));
        //参数1:圆弧起始角度,单位为度;参数2:圆弧扫过的角度,顺时针方向,单位为度;
        //参数3:如果为True时,在绘制圆弧时将圆心包括在内,通常用来绘制扇形;参数4:画笔
        canvas.drawArc(oval,0,360*progress/max,false,paint);
        //步骤4:展示进度文字
        paint.setStrokeWidth(0);
        paint.setColor(mProgressColor);
        paint.setTextSize(mProgressTextSize);
        //设置字体
        paint.setTypeface(Typeface.DEFAULT_BOLD);
        int percent=(int)(((float)progress/(float)max)*100);
        //中间的进度百分比,先转换成float在进行除法运算,不然都为0
        float textWidth=paint.measureText(percent+"%");
        if(percent>0){
            //画出进度百分比
            canvas.drawText(percent+"%",centerX-textWidth/2,
            (float)(centerY + radius-mDistanceOFbg*6),paint);
        }
    }
    /**设置进度,此为线程安全控件,由于考虑多线程的问题,需要线程同步*/
    public synchronized void setProcess(int process){
        if(process<0){
            throw new IllegalArgumentException("progress not less than 0");
        }
        if(process>max){
            process=max;
        }
        if(process<=max){
            this.progress=process;
            //重绘图片
            postInvalidate();
        }
    }
```

```
101    public synchronized void setMax(int max){
102        if(max<0){
103            throw new IllegalArgumentException("max not less than 0");
104        }
105        this.max=max;
106    }
107 }
```

代码说明：

- 第 14～23 行代码实现了 Button 控件的三个构造方法，并在构造方法中对布局控件进行初始化赋值。
- 第 39～41 行代码使用了 DensityUtil 工具类中的单位转换将 dip 转换为 px，该工具类在第 4 章讲解过，直接复制过来用即可。
- 第 45～86 行的 onDraw()方法，用于实现绘图功能，我们主要绘制三部分：最外层的圆圈；内层的圆形进度条；圆心的进度文字。首先在第 48～50 行代码计算出圆心的位置，然后在第 52～56 行代码画出外层的圆圈；在第 58～71 行代码画出外层的进度条；在第 73～85 行代码展示进度文字。
- 第 88～106 行代码定义了两个方法，分别是设置进度条的进度和设置进度条的最大值。需要注意的是，考虑到多线程安全问题，这两个方法都设置为 synchronized 线程同步。

短信备份的自定义控件编写完成后，接下来在布局文件中使用该控件，具体效果如图 9-11 所示。

图 9-11 所示短信备份界面对应的布局文件如【文件 9-15】所示。

图 9-11 短信备份界面

【文件 9-15】activity_smsbackup.xml

```
<?xml version="1.0" encoding="utf-8"?>
<LinearLayout xmlns:android="http://schemas.android.com/apk/res/android"
    xmlns:itcast="http://schemas.android.com/apk/res/cn.itcast.mobliesafe"
    android:layout_width="match_parent"
    android:layout_height="match_parent"
    android:orientation="vertical"
    android:background="@color/graye5">
    <include layout="@layout/titlebar"/>
    <LinearLayout
        android:layout_width="match_parent"
        android:layout_height="match_parent"
        android:gravity="center">
        <cn.itcast.mobliesafe.chapter09.widget.MyCircleProgress
            android:id="@+id/mcp_smsbackup"
            android:layout_width="wrap_content"
```

```xml
            android:layout_height="wrap_content"
            android:layout_gravity="center"
            android:gravity="center"
            android:text="一键备份"
            android:textColor="@color/white"
            itcast:circleColor="@color/bright_red"
            itcast:progressColor="@color/light_pink"
            android:background="@drawable/smsbackup_button_bg"/>
    </LinearLayout>
</LinearLayout>
```

在上述代码中使用了自定义控件，在布局中开始节点和结束节点设置为该自定义类的全路径名 cn.itcast.mobliesafe.chapter09.widget.MyCircleProgress，节点中的 itcast:circleColor 属性为自定义控件中圆心的颜色，itcast:progressColor 属性为进度百分比的颜色。

9.4.4 短信备份逻辑

当短信备份功能所使用的工具类和布局都定义好之后，下面开始编写短信备份主逻辑代码，主要控制短信的备份和取消、备份进度、文字的显示以及备份状态的显示，具体代码如【文件 9-16】所示。

【文件 9-16】SMSBackupActivity.java

```java
1  /**短信备份*/
2  public class SMSBackupActivity extends Activity implements OnClickListener{
3      private MyCircleProgress mProgressButton;
4      /**标识符，用来标识备份状态的*/
5      private boolean flag=false;
6      private SmsBackUpUtils smsBackUpUtils;
7      private static final int CHANGE_BUTTON_TEXT=100;
8      private Handler handler=new Handler(){
9          public void handleMessage(android.os.Message msg){
10             switch(msg.what){
11             case CHANGE_BUTTON_TEXT:
12                 mProgressButton.setText("一键备份");
13                 break;
14             }
15         };
16     };
17     @Override
18     protected void onCreate(Bundle savedInstanceState){
19         super.onCreate(savedInstanceState);
20         requestWindowFeature(Window.FEATURE_NO_TITLE);
```

```java
21      setContentView(R.layout.activity_smsbackup);
22      smsBackUpUtils=new SmsBackUpUtils();
23      initView();
24  }
25  private void initView(){
26      findViewById(R.id.rl_titlebar).setBackgroundColor(getResources().getColor(
27      R.color.bright_red));
28      ImageView mLeftImgv=(ImageView) findViewById(R.id.imgv_leftbtn);
29      ((TextView) findViewById(R.id.tv_title)).setText("短信备份");
30      mLeftImgv.setOnClickListener(this);
31      mLeftImgv.setImageResource(R.drawable.back);
32      mProgressButton=(MyCircleProgress) findViewById(R.id.mcp_smsbackup);
33      mProgressButton.setOnClickListener(this);
34  }
35  @Override
36  protected void onDestroy(){
37      flag=false;
38      smsBackUpUtils.setFlag(flag);
39      super.onDestroy();
40  }
41  @Override
42  public void onClick(View v){
43      switch (v.getId()){
44      case R.id.imgv_leftbtn:
45          finish();
46          break;
47      case R.id.mcp_smsbackup:
48          if(flag){
49              flag=false;
50              mProgressButton.setText("一键备份");
51          }else{
52              flag=true;
53              mProgressButton.setText("取消备份");
54          }
55          smsBackUpUtils.setFlag(flag);
56          new Thread(){
57              public void run(){
58                  try{
59                      boolean backUpSms=smsBackUpUtils.backUpSms(
```

```
60                    SMSBackupActivity.this, new BackupStatusCallback(){
61                        @Override
62                        public void onSmsBackup(int process){
63                            mProgressButton.setProcess(process);
64                        }
65                        @Override
66                        public void beforeSmsBackup(int size){
67                            if(size<=0){
68                                flag=false;
69                                smsBackUpUtils.setFlag(flag);
70                                UIUtils.showToast(SMSBackupActivity.this,
71                                "您还没有短信!");
72                                handler.sendEmptyMessage(CHANGE_BUTTON_TEXT);
73                            }else{
74                                mProgressButton.setMax(size);
75                            }
76                        }
77                    });
78                    if(backUpSms){
79                        UIUtils.showToast(SMSBackupActivity.this,"备份成功");
80                    } else {
81                        UIUtils.showToast(SMSBackupActivity.this,"备份失败");
82                    }
83                }catch(FileNotFoundException e){
84                    e.printStackTrace();
85                    UIUtils.showToast(SMSBackupActivity.this,"文件生成失败");
86                }catch(IllegalStateException e){
87                    e.printStackTrace();
88                    UIUtils.showToast(SMSBackupActivity.this,
89                    "SD卡不可用或SD卡内存不足");
90                }catch(IOException e){
91                    e.printStackTrace();
92                    UIUtils.showToast(SMSBackupActivity.this,"读写错误");
93                }
94            };
95        }.start();
96        break;
97    }
98 }
99}
```

代码说明：

- 第 25～34 行的 initView()方法用于获得自定义控件对象并设置点击监听。需要通过自定义控件的对象设置进度条和按钮文字的变化。
- 第 42～98 行的 onClick()方法用于设置界面中短信备份按钮的点击监听事件。
- 第 48～55 行代码用到了一个全局变量 boolean 类型的 flag，如果 flag 为 true 说明正在进行短信备份操作，取消正在备份的短信，将按钮文字改为一键备份并设置 flag 为 false；如果 flag 为 false 说明正在备份中，将按钮文字改为取消备份并设置 flag 为 true。当 flag 的值改变之后，通过 smsBackUpUtils.setFlag(flag)方法将重新设置后的 flag 的值传递给短信备份的工具类。
- 第 56～95 行代码在开启子线程中调用短信备份工具类 smsBackUpUtils 的 backUpSms()方法开始进行短信备份。并实现该类中定义回调接口的两个方法，其中在 onSmsBackup()方法中进行进度条的设置，在 beforeSmsBackup()方法中根据参数中得短信条数进行判断，如果没有短信，则弹出 Toast 提示并发送 Handler 将按钮文字改为"一键备份"。需要注意的是，只要当 flag 设置为 true 时才会真正开始短信备份的操作，因为 backUpSms()方法中开始序列化短信时有一个条件就是 flag 的值为 true，这个条件在【文件 9-11】的第 22 行代码中设置。

至此，短信备份功能全部完成，在高级工具模块点击"短信备份"按钮进入该界面，点击中间的备份按钮，会看到按钮周围的进度条开始变化。此时打开 DDMS，点击 File Explorer 打开文件浏览器，然后打开 SD 卡，若能找到 backup.xml 文件，说明短信备份已经成功，由于该备份的文件已经加密，所以打开看到的会是乱码，这是正常现象。

9.4.5 Toast 封装

在编写程序时，很多地方都需要用到 Toast 进行提示，为此可以对 Toast 进行一个封装，方便后面使用，具体代码如【文件 9-17】所示。

【文件 9-17】UIUtils.java

```
1  public class UIUtils{
2      public static void showToast(final Activity context,final String msg){
3          if("main".equals(Thread.currentThread().getName())){
4              Toast.makeText(context,msg,1).show();
5          }else{
6              context.runOnUiThread(new Runnable(){
7                  @Override
8                  public void run(){
9                      Toast.makeText(context,msg,1).show();
10                 }
11             });
12         }
13     }
14 }
```

在上述代码中,定义了一个静态方法 showToast(),接收两个参数为上下文 context 和需要显示的消息 msg。如果当前线程为主线程则直接弹出 Toast 提示,否则让 Toast 运行在主线程中。这样在使用 Toast 时,直接使用"类名.方法名"调用即可,不仅使用方便而且大大提高程序的健壮性。

9.5 短信还原

9.5.1 短信还原工具类

短信还原功能和短信备份功能是相对应的,短信还原功能是将手机中已删除的短信记录全部还原,由于备份短信时使用的是 XML 格式保存的,因此在还原时,同样需要使用 XML 解析器将已备份的短信还原,具体代码如【文件 9-18】所示。

【文件 9-18】SmsReducitionUtils.java

```
1  /**
2   * 短信还原的工具类
3   */
4  public class SmsReducitionUtils{
5      public interface SmsReducitionCallBack{
6          /**
7           * 在短信还原之前调用的方法
8           * @param size
9           * 总的短信的个数
10          */
11         public void beforeSmsReduction(int size);
12         /**
13          * 当 sms 短信还原过程中调用的方法
14          * @param process 当前的进度
15          */
16         public void onSmsReduction(int process);
17     }
18     private boolean flag=true;
19     public void setFlag(boolean flag){
20         this.flag=flag;
21     }
22     public boolean reductionSms(Activity context,SmsReducitionCallBack callBack)
23         throws XmlPullParserException,IOException{
24         File file=new File(Environment.getExternalStorageDirectory(),"backup.xml");
25         if(file.exists()){
26             FileInputStream is=new FileInputStream(file);
```

```java
27      XmlPullParser parser=Xml.newPullParser();
28      parser.setInput(is,"utf-8");
29      SmsInfo smsInfo=null;
30      int eventType=parser.getEventType();
31      Integer max=null;
32      int progress=0;
33      ContentResolver resolver=context.getContentResolver();
34      Uri uri=Uri.parse("content://sms/");
35      while(eventType!=XmlPullParser.END_DOCUMENT & flag){
36          switch(eventType){
37          //一个节点的开始
38          case XmlPullParser.START_TAG:
39              if("smss".equals(parser.getName())){
40                  String maxStr=parser.getAttributeValue(0);
41                  max=new Integer(maxStr);
42                  callBack.beforeSmsReducition(max);
43              }else if("sms".equals(parser.getName())){
44                  smsInfo=new SmsInfo();
45              }else if("body".equals(parser.getName())){
46                  try{
47                      smsInfo.body = Crypto.decrypt("123", parser.nextText());
48                  }catch(Exception e){
49                      e.printStackTrace();
50                      //此条短信还原失败
51                      smsInfo.body="短信还原失败";
52                  }
53              }else if("address".equals(parser.getName())){
54                  smsInfo.address=parser.nextText();
55              }else if("type".equals(parser.getName())){
56                  smsInfo.type=parser.nextText();
57              }else if("date".equals(parser.getName())){
58                  smsInfo.date=parser.nextText();
59              }
60              break;
61          //一个节点的结束
62          case XmlPullParser.END_TAG:
63              if("sms".equals(parser.getName())){
64                  //向短信数据库中插入一条数据
65                  ContentValues values=new ContentValues();
```

```
66                  values.put("address",smsInfo.address);
67                  values.put("type",smsInfo.type);
68                  values.put("date",smsInfo.date);
69                  values.put("body",smsInfo.body);
70                  resolver.insert(uri,values);
71                  smsInfo=null;
72                  progress++;
73                  callBack.onSmsReducition(progress);
74              }
75              break;
76          }
77          //得到下一个节点的事件类型,此行代码一定不能忘否则会成死循环
78          eventType=parser.next();
79      }
80      //防止出现在备份未完成的情况下,还原短信
81      if(eventType==XmlPullParser.END_DOCUMENT & max!=null){
82          if(progress<max){
83              callBack.onSmsReducition(max);
84          }
85      }
86  }else{
87      //如果backup.xml文件不存在,则说明没有备份短信
88      UIUtils.showToast(context,"您还没有备份短信!");
89  }
90  return flag;
91  }
92 }
```

代码说明:

- 第 5~17 行代码定义了一个回调接口 SmsReducitionCallBack,该接口中有两个方法 beforeSmsReducition()和 onSmsReducition(),其中 beforeSmsReducition()方法在第 42 行短信还原之前调用,获取将要还原的短信总条数;onSmsReducition()方法在第 73 行短信还原过程中调用,得到当前还原的进度。该回调接口在 SMSReducitionActivity 界面短信还原按钮监听中调用。
- 第 18~21 行代码定义了一个 boolean 值变量 flag,并定义了 setFlag()方法为 flag 赋值。该变量主要控制短信还原与取消。
- 第 22~91 行代码为短信还原的主逻辑。首先在第 24 行得到之前备份的文件,文件名必须和备份时定义的名称一致。如果文件存在,则使用 XmlPullParser 进行 XML 文件的解析。在第 63~74 行代码中获取一个节点的数据就向系统短信数据库中写入一条信息。

9.5.2 短信的实体类

在还原短信时,需要定义一个实体类,用于存储短信数据,该实体类中包含的字段有短信内容、短信日期、短信类型、短信地址,具体代码如【文件9-19】所示。

【文件9-19】SmsInfo.java

```
1 public class SmsInfo{
2     public String body;
3     public String date;
4     public String type;
5     public String address;
6 }
```

9.5.3 短信还原界面

短信还原界面与短信备份界面是一样的,都使用了之前自定义的红色按钮(MyCircleProgress.java,这里不再贴出该类的代码),短信还原界面如图9-12所示。

从图9-12可以看出,短信还原界面与短信备份界面唯一不同的地方就是中间按钮的文字,该界面对应的布局文件如【文件9-20】所示。

【文件9-20】activity_reduction.xml

图9-12 短信还原界面

```
<?xml version="1.0" encoding="utf-8"?>
<LinearLayout xmlns:android="http://schemas.android.
   com/apk/res/android"
   xmlns:itcast="http://schemas.android.com/apk/res/cn.itcast.mobliesafe"
   android:layout_width="match_parent"
   android:layout_height="match_parent"
   android:background="@color/graye5"
   android:orientation="vertical" >
   <include layout="@layout/titlebar"/>
   <LinearLayout
      android:layout_width="match_parent"
      android:layout_height="match_parent"
      android:gravity="center" >
      <cn.itcast.mobliesafe.chapter09.widget.MyCircleProgress
         android:id="@+id/mcp_reduction"
         android:layout_width="wrap_content"
         android:layout_height="wrap_content"
         android:layout_gravity="center"
         android:background="@drawable/smsbackup_buuton_bg"
```

```
                android:gravity="center"
                android:text="一键还原"
                android:textColor="@color/white"
                itcast:circleColor="@color/bright_red"
                itcast:progressColor="@color/light_pink"/>
    </LinearLayout>
</LinearLayout>
```

9.5.4 短信还原逻辑

当短信还原功能使用的工具类和布局都定义好之后,接下来编写短信还原主逻辑代码,控制短信的还原、取消以及还原进度等,具体代码如【文件 9-21】所示。

【文件 9-21】SMSReducitionActivity.java

```
1  /**短信还原*/
2  public class SMSReducitionActivity extends Activity implements OnClickListener{
3      private MyCircleProgress mProgressButton;
4      private boolean flag=false;
5      private SmsReducitionUtils smsReducitionUtils;
6      @Override
7      protected void onCreate(Bundle savedInstanceState){
8          super.onCreate(savedInstanceState);
9          requestWindowFeature(Window.FEATURE_NO_TITLE);
10         setContentView(R.layout.activity_reduction);
11         initView();
12         smsReducitionUtils=new SmsReducitionUtils();
13     }
14     private void initView(){
15         findViewById(R.id.rl_titlebar).setBackgroundColor(
16             getResources().getColor(R.color.bright_red));
17         ImageView mLeftImgv=(ImageView) findViewById(R.id.imgv_leftbtn);
18         ((TextView) findViewById(R.id.tv_title)).setText("短信还原");
19         mLeftImgv.setOnClickListener(this);
20         mLeftImgv.setImageResource(R.drawable.back);
21         mProgressButton=(MyCircleProgress) findViewById(R.id.mcp_reduction);
22         mProgressButton.setOnClickListener(this);
23     }
24     @Override
25     protected void onDestroy(){
26         flag=false;
27         smsReducitionUtils.setFlag(flag);
```

```java
28      super.onDestroy();
29  }
30  @Override
31  public void onClick(View v){
32      switch (v.getId()){
33      case R.id.imgv_leftbtn:
34          finish();
35          break;
36      case R.id.mcp_reducition:
37          if(flag){
38              flag=false;
39              mProgressButton.setText("一键还原");
40          }else{
41              flag=true;
42              mProgressButton.setText("取消还原");
43          }
44          smsReducitionUtils.setFlag(flag);
45          new Thread(){
46              public void run(){
47                  try{
48                      smsReducitionUtils.reducitionSms(SMSReducitionActivity.this,
49                      new SmsReducitionCallBack(){
50                          @Override
51                          public void onSmsReducition(int process){
52                              mProgressButton.setProcess(process);
53                          }
54                          @Override
55                          public void beforeSmsReducition(int size){
56                              mProgressButton.setMax(size);
57                          }
58                      });
59                  }catch (XmlPullParserException e){
60                      e.printStackTrace();
61                      UIUtils.showToast(SMSReducitionActivity.this,"文件格式错误");
62                  }catch (IOException e){
63                      e.printStackTrace();
64                      UIUtils.showToast(SMSReducitionActivity.this,"读写错误");
65                  }
66              }
```

```
67            }.start();
68            break;
69        }
70    }
71 }
```

代码说明：

- 第 25～29 行的 onDestory()方法是在界面关闭时执行，在该方法中将 flag 赋值为 false，然后通过 SmsReducitionUtils 类的 setFlag()方法控制停止短信还原，也就是说当界面关闭了，短信还原功能停止。
- 第 31～70 行的 onClick()方法用于监听界面的点击事件，这段代码逻辑与短信备份的一样，首先设置 flag 的值来控制按钮文字的显示和是否进行短信还原。其中第 48～57 行代码实现了 SmsReducitionUtils 类中的回调方法，在 onSmsReducition()方法中设置进度条进度，在 beforeSmsReducition()方法中设置需要还原的短信的总个数。

至此，短信还原功能已经完成。先使用短信备份功能将短信保存在本地并删除短信，然后点击"一键还原"按钮会看到进度条开始变化，当进度条完成之后，打开系统短信界面，会看到之前备份的短信显示在界面上，证明短信还原功能已经完成。

9.6 程 序 锁

程序锁功能主要是将加锁的程序信息存入数据库，当程序锁服务打开时，后台会运行一个服务检查当前打开的程序，如果程序在数据库中说明是加锁的，此时会弹出输入密码界面，只有密码输入正确才能进入应用（这个密码就是手机防盗模块中的防盗密码），本节将针对程序锁进行详细讲解。

9.6.1 创建数据库

在编写程序锁逻辑之前，先创建存储已加锁应用的数据库以及数据库操作类，创建数据库代码如【文件 9-22】所示。

【文件 9-22】AppLockOpenHelper.java

```
1  public class AppLockOpenHelper extends SQLiteOpenHelper{
2      public AppLockOpenHelper(Context context) {
3          super(context,"applock.db",null,1);
4      }
5      @Override
6      public void onCreate(SQLiteDatabase db){
7          db.execSQL("create table applock(id integer primary key autoincrement,
8          packagename  varchar(20))");
9      }
10     @Override
11     public void onUpgrade(SQLiteDatabase db,int oldVersion,int newVersion){
12     }
13 }
```

在上述代码中，创建了一个 applock.db 数据库，在该数据库中定义了一个 applock 表，表中有两个属性：id 和 packagename。

9.6.2 数据库操作类

数据库创建完成之后，需要对数据库进行增删改查的操作，接下来就创建该数据库的操作类，具体代码如【文件 9-23】所示。

【文件 9-23】AppLockDao.java

```
1  /** 程序锁数据库操作逻辑类 */
2  public class AppLockDao {
3     private Context context;
4     private AppLockOpenHelper openHelper;
5     private Uri uri=Uri.parse("content://com.itcast.mobilesafe.applock");
6     public AppLockDao(Context context){
7        this.context=context;
8        openHelper=new AppLockOpenHelper(context);
9     }
10    /**
11     * 添加一条数据
12     * @return
13     */
14    public boolean insert(String packagename){
15       SQLiteDatabase db=openHelper.getWritableDatabase();
16       ContentValues values=new ContentValues();
17       values.put("packagename",packagename);
18       long rowid=db.insert("applock",null,values);
19       if(rowid==-1) //插入不成功
20          return false;
21       else{// 插入成功
22          context.getContentResolver().notifyChange(uri, null);
23          return true;
24       }
25    }
26    /**
27     * 删除一条数据
28     * @param packagename
29     * @return
30     */
31    public boolean delete(String packagename){
32       SQLiteDatabase db = openHelper.getWritableDatabase();
```

```
33      int rownum=db.delete("applock","packagename=?",newString[] { packagename });
34      if(rownum==0){
35          return false;
36      }else{
37          context.getContentResolver().notifyChange(uri,null);
38          return true;
39      }
40  }
41  /**
42   * 查询某个包名是否存在
43   * @param packagename
44   * @return
45   */
46  public boolean find(String packagename){
47      SQLiteDatabase db=openHelper.getReadableDatabase();
48      Cursor cursor=db.query("applock",null,"packagename=?",
49      new String[] { packagename },null,null,null);
50      if(cursor.moveToNext()){
51          cursor.close();
52          db.close();
53          return true;
54      }else{
55          cursor.close();
56          return false;
57      }
58  }
59  /**
60   * 查询表中所有的包名
61   * @return
62   */
63  public List<String> findAll(){
64      SQLiteDatabase db=openHelper.getReadableDatabase();
65      Cursor cursor=db.query("applock",null,null,null,null,null,null);
66      List<String> packages=new ArrayList<String>();
67      while(cursor.moveToNext()){
68          String string = cursor.getString(cursor.getColumnIndex("packagename"));
69          packages.add(string);
70      }
71      return packages;
```

```
72    }
73 }
```

代码说明：

- 第 14~25 行代码为添加数据，参数接收应用程序的包名，如果数据添加成功，则通知内容观察者内容发生变化。
- 第 31~40 行代码为删除数据，同样地如果数据删除成功通知内容观察者。
- 第 63~72 行代码查询数据库中所有的数据，返回 List 集合，集合中包含应用数据的包名。

需要注意的是，在添加和删除方法中都使用了 ContentResolver 暴露数据，而注册 ContentResolver 的 uri 是在第 5 行代码所自定义的 content://com.itcast.mobilesafe.applock。然后会在已加锁界面、未加锁界面、程序锁服务中使用该 uri 进行注册内容观察者，如果使用数据库进行增加和删除功能导致数据变化时会通知注册了该 uri 的内容观察者以便做相应地操作，该操作在 9.6.7 小节中会详细讲解。

9.6.3 获取所有应用工具类

要实现程序的加锁功能，首先要获取到手机上的所有应用程序，获取方法在前面几章已经讲解过，这里不再过多解释，具体代码如【文件 9-24】所示。

【文件 9-24】AppInfoParser.java

```
1  /**
2   * 工具类用来获取应用信息，此类重复
3   */
4  public class AppInfoParser{
5      /**
6       * 获取手机里面的所有的应用程序
7       * @param context 上下文
8       * @return
9       */
10     public static List<AppInfo> getAppInfos(Context context){
11         //得到一个包管理器
12         PackageManager pm=context.getPackageManager();
13         List<PackageInfo> packInfos=pm.getInstalledPackages(0);
14         List<AppInfo> appinfos=new ArrayList<AppInfo>();
15         for(PackageInfo packInfo:packInfos){
16             AppInfo appinfo=new AppInfo();
17             String packname=packInfo.packageName;
18             appinfo.packageName=packname;
19             Drawable icon=packInfo.applicationInfo.loadIcon(pm);
20             appinfo.icon=icon;
21             String appname=packInfo.applicationInfo.loadLabel(pm).toString();
22             appinfo.appName=appname;
```

```
23          //应用程序 APK 包的路径
24          String apkpath=packInfo.applicationInfo.sourceDir;
25          appinfo.apkPath=apkpath;
26          appinfos.add(appinfo);
27          appinfo=null;
28      }
29      return appinfos;
30  }
31 }
```

9.6.4 应用的实体类

为了方便对应用进行操作，需要定义一个应用程序的实体类，该类中包含包名、图标、应用名、应用路径、是否加锁，具体代码如【文件 9-25】所示。

【文件 9-25】AppInfo.java

```
1  public class AppInfo{
2      /** 应用程序包名 */
3      public String packageName;
4      /** 应用程序图标 */
5      public Drawable icon;
6      /** 应用程序名称 */
7      public String appName;
8      /** 应用程序路径 */
9      public String apkPath;
10     /** 应用程序是否加锁 */
11     public boolean isLock;
12 }
```

9.6.5 程序锁界面

程序锁界面主要包含两个 Fragment，分别用于展示未加锁界面和已加锁程序界面，并使用 ViewPager 控件使这两个页面可以滑动切换，具体如图 9-13 所示。

图 9-13 所示页面中显示了两个按钮，分别是未加锁和已加锁，点击这两个按钮会切换到不同页面，在未加锁的下方有一根红色线条，当 ViewPager 滑动到某一个页面时文字就会变成红色，并且线条也会滑动到相应按钮下面。程序锁界面布局如【文件 9-26】所示。

【文件 9-26】activity_applock.xml

```
<?xml version="1.0" encoding="utf-8"?>
<LinearLayout
```

图 9-13 程序锁界面

```xml
    xmlns:android="http://schemas.android.com/apk/res/android"
    android:layout_width="match_parent"
    android:layout_height="match_parent"
    android:orientation="vertical">
    <include layout="@layout/titlebar"/>
    <LinearLayout
        android:layout_width="match_parent"
        android:layout_height="35dp"
        android:orientation="horizontal"
        android:background="@color/graye5"
        android:gravity="center_vertical">
        <TextView
            android:id="@+id/tv_unlock"
            style="@style/zero_widthwrapcontent"
            android:textSize="16sp"
            android:gravity="center"
            android:padding="5dp"
            android:textColor="@color/bright_red"
            android:text="未加锁"/>
        <TextView
            android:id="@+id/tv_lock"
            style="@style/zero_widthwrapcontent"
            android:textSize="16sp"
            android:gravity="center"
            android:padding="5dp"
            android:text="已加锁"/>
    </LinearLayout>
    <LinearLayout
        android:layout_width="match_parent"
        android:layout_height="wrap_content"
        android:background="@color/graye5"
        android:orientation="horizontal">
        <View
            android:id="@+id/view_slide_unlock"
            style="@style/zero_widthwrapcontent"
            android:layout_height="3dp"
            android:background="@drawable/slide_view"/>
        <View
            android:id="@+id/view_slide_lock"
```

```
            style="@style/zero_widthwrapcontent"
            android:layout_height="3dp"
            android:layout_gravity="bottom"/>
    </LinearLayout>
    <android.support.v4.view.ViewPager
        android:id="@+id/vp_applock"
        android:layout_width="match_parent"
        android:layout_height="match_parent" />
</LinearLayout>
```

在上述布局中，最顶端定义了一个 LinearLayout 用于显示两个 TextView，分别显示未加锁和已加锁；下方定义了一个 LinearLayout 用于显示红色线条；在页面最下方定义了 ViewPager 用于进行页面的滑动切换。其中 TextView 控件以及 View 控件都使用了一个 zero_widthwrapcontent 样式，用于控制宽、高、权重的展示形式，具体代码如【文件 9-27】所示。

【文件 9-27】res/values/styles.xml

```
<!-- 宽度为 0dp, 高度为包裹内容, 权重为 1 -->
<style name="zero_widthwrapcontent">
    <item name="android:layout_width">0dp</item>
    <item name="android:layout_height">wrap_content</item>
    <item name="android:layout_weight">1</item>
</style>
```

9.6.6 程序锁逻辑

程序锁的主要功能就是控制两个 Fragment 的切换，对于未加锁和已加锁应用的操作都在 Fragment 中进行，具体代码如【文件 9-28】所示。

【文件 9-28】AppLockActivity.java

```
1  /** 程序锁 */
2  public class AppLockActivity extends FragmentActivity implements
3  OnClickListener{
4      private ViewPager mAppViewPager;
5      List<Fragment> mFragments=new ArrayList<Fragment>();
6      private TextView mLockTV;
7      private TextView mUnLockTV;
8      private View slideLockView;
9      private View slideUnLockView;
10     @Override
11     protected void onCreate(Bundle savedInstanceState){
12         super.onCreate(savedInstanceState);
```

```java
13      requestWindowFeature(Window.FEATURE_NO_TITLE);
14      setContentView(R.layout.activity_applock);
15      initView();
16      initListener();
17   }
18   private void initListener(){
19      mAppViewPager.setOnPageChangeListener(new OnPageChangeListener(){
20         @Override
21         public void onPageSelected(int arg0){
22            if(arg0==0){
23               slideUnLockView.setBackgroundResource(R.drawable.slide_view);
24               slideLockView.setBackgroundColor(getResources().getColor(
25               R.color.transparent));
26               //未加锁
27               mLockTV.setTextColor(getResources().getColor(R.color.black));
28               mUnLockTV.setTextColor(getResources().getColor(
29               R.color.bright_red));
30            }else{
31               slideLockView.setBackgroundResource(R.drawable.slide_view);
32               slideUnLockView.setBackgroundColor(getResources().getColor(
33               R.color.transparent));
34               //已加锁
35               mLockTV.setTextColor(getResources().getColor(
36               R.color.bright_red));
37               mUnLockTV.setTextColor(getResources().getColor(R.color.black));
38            }
39         }
40         @Override
41         public void onPageScrolled(int arg0,float arg1,int arg2){
42         }
43         @Override
44         public void onPageScrollStateChanged(int arg0){
45         }
46      });
47   }
48   private void initView(){
49      findViewById(R.id.rl_titlebar).setBackgroundColor(
50      getResources().getColor(R.color.bright_red));
51      ImageView mLeftImgv=(ImageView) findViewById(R.id.imgv_leftbtn);
```

```
52      ((TextView) findViewById(R.id.tv_title)).setText("程序锁");
53      mLeftImgv.setOnClickListener(this);
54      mLeftImgv.setImageResource(R.drawable.back);
55      mAppViewPager=(ViewPager) findViewById(R.id.vp_applock);
56      mLockTV=(TextView) findViewById(R.id.tv_lock);
57      mUnLockTV=(TextView) findViewById(R.id.tv_unlock);
58      mLockTV.setOnClickListener(this);
59      mUnLockTV.setOnClickListener(this);
60      slideLockView=findViewById(R.id.view_slide_lock);
61      slideUnLockView=findViewById(R.id.view_slide_unlock);
62      AppUnLockFragment unLock=new AppUnLockFragment();
63      AppLockFragment lock=new AppLockFragment();
64      mFragments.add(unLock);
65      mFragments.add(lock);
66      mAppViewPager.setAdapter(new MyAdapter(getSupportFragmentManager()));
67  }
68  @Override
69  public void onClick(View v){
70      switch(v.getId()){
71      case R.id.imgv_leftbtn:
72          finish();
73          break;
74      case R.id.tv_lock:
75          mAppViewPager.setCurrentItem(1);
76          break;
77      case R.id.tv_unlock:
78          mAppViewPager.setCurrentItem(0);
79          break;
80      }
81  }
82  class MyAdapter extends FragmentPagerAdapter{
83      public MyAdapter(FragmentManager fm){
84          super(fm);
85      }
86      @Override
87      public android.support.v4.app.Fragment getItem(int arg0){
88          return mFragments.get(arg0);
89      }
90      @Override
```

```
91        public int getCount() {
92            return mFragments.size();
93        }
94    }
95 }
```

代码说明：
- 第 5 行代码定义了一个 ArrayList 集合，该集合主要存放两个 Fragment。
- 第 18～47 行代码判断当前页面是哪一个页面，用于控制文字的颜色和线条的位置。
- 第 48～67 行代码初始化页面控件，并实例化两个 Fragment，将 Fragment 的引用加入集合中。
- 第 69～81 行代码为两个按钮的点击事件，使用 ViewPager 的 setCurrentItem()方法进行页面的跳转。
- 第 82～94 行的创建了 ViewPager 的 Adapter，用于控制 ViewPager 的页面个数与加载条目。

9.6.7 加锁与未加锁功能

当进入程序锁模块时默认显示的是未加锁界面，当第一次打开该模块时会将手机中安装的所有应用信息都显示在界面中，当点击某一个应用时会将该应用从未加锁界面删除并添加到已加锁界面中，并且被删除的条目会有一个滑动删除的动画。接下来分步骤实现这个程序锁的 Fragment，具体如下：

1. **未加锁界面**

在 Fragment 中有两个界面，一个是未加锁界面，另一个是已加锁界面，这两个界面中都只有一个 TextView 控件和一个 ListView 控件，分别用于显示加锁（未加锁）应用的个数以及加锁（未加锁）的程序列表。由于这两个布局比较简单，因此不进行图形化界面展示，未加锁界面代码如【文件 9-29】所示。

【文件 9-29】fragment_appunlock.xml

```xml
<?xml version="1.0" encoding="utf-8"?>
<LinearLayout xmlns:android="http://schemas.android.com/apk/res/android"
    android:layout_width="match_parent"
    android:layout_height="match_parent"
    android:orientation="vertical" >
    <TextView
        android:id="@+id/tv_unlock"
        android:layout_width="match_parent"
        android:layout_height="wrap_content"
        android:padding="5dp"
        android:background="@color/graye5"/>
    <ListView
        android:id="@+id/lv_unlock"
```

```
        android:layout_width="match_parent"
        android:layout_height="wrap_content"/>
</LinearLayout>
```

2. 已加锁界面

已加锁界面和未加锁界面是一样的，都使用了一个 TextView 控件和一个 ListView 控件，已加锁界面代码如【文件 9-30】所示。

【文件 9-30】fragment_applock.xml

```xml
<?xml version="1.0" encoding="utf-8"?>
<LinearLayout xmlns:android="http://schemas.android.com/apk/res/android"
    android:layout_width="match_parent"
    android:layout_height="match_parent"
    android:orientation="vertical">
    <TextView
        android:id="@+id/tv_lock"
        android:layout_width="match_parent"
        android:layout_height="wrap_content"
        android:padding="5dp"
        android:background="@color/graye5"/>
    <ListView
        android:id="@+id/lv_lock"
        android:layout_width="match_parent"
        android:layout_height="wrap_content"/>
</LinearLayout>
```

3. 未加锁逻辑代码

当进入程序锁模块时默认显示的就是未加锁界面，如果是第一次打开该界面就显示在手机上安装的所有应用，当在该界面将某个应用程序加锁时，就将该应用从未加锁界面移除。未加锁逻辑代码如【文件 9-31】所示。

【文件 9-31】AppUnLockFragment.java

```java
1  public class AppUnLockFragment extends Fragment{
2      private TextView mUnLockTV;
3      private ListView mUnLockLV;
4      List<AppInfo> unlockApps = new ArrayList<AppInfo>();
5      private AppLockAdapter adapter;
6      private AppLockDao dao;
7      private Uri uri = Uri.parse("content://com.itcast.mobilesafe.applock");
8      private List<AppInfo> appInfos;
9      private Handler mhandler=new Handler(){
10         public void handleMessage(android.os.Message msg){
```

```java
11       switch (msg.what){
12       case 100:
13           unlockApps.clear();
14           unlockApps.addAll(((List<AppInfo>)msg.obj));
15           if(adapter==null){
16               adapter=new AppLockAdapter(unlockApps, getActivity());
17               mUnLockLV.setAdapter(adapter);
18           }else{
19               adapter.notifyDataSetChanged();
20           }
21           mUnLockTV.setText("未加锁应用"+unlockApps.size()+"个");
22           break;
23       }
24   }
25 };
26 @Override
27 public View onCreateView(LayoutInflater inflater,ViewGroup container,
28 Bundle savedInstanceState){
29     View view=inflater.inflate(R.layout.fragment_appunlock,null);
30     mUnLockTV=(TextView) view.findViewById(R.id.tv_unlock);
31     mUnLockLV=(ListView) view.findViewById(R.id.lv_unlock);
32     return view;
33 }
34 @Override
35 public void onResume(){
36     dao=new AppLockDao(getActivity());
37     appInfos=AppInfoParser.getAppInfos(getActivity());
38     fillData();
39     initListener();
40     super.onResume();
41     getActivity().getContentResolver().registerContentObserver(uri, true,
42     new ContentObserver(new Handler()){
43         @Override
44         public void onChange(boolean selfChange){
45             fillData();
46         }
47     });
48 }
49 public void fillData(){
```

```
50      final List<AppInfo> aInfos=new ArrayList<AppInfo>();
51      new Thread(){
52          public void run(){
53              for(AppInfo info:appInfos){
54                  if(!dao.find(info.packageName)){
55                      //未加锁
56                      info.isLock=false;
57                      aInfos.add(info);
58                  }
59              }
60              Message msg=new Message();
61              msg.obj=aInfos;
62              msg.what=100;
63              mhandler.sendMessage(msg);
64          };
65      }.start();
66  }
67  private void initListener(){
68      mUnLockLV.setOnItemClickListener(new OnItemClickListener(){
69          @Override
70          public void onItemClick(AdapterView<?> parent,View view,
71          final int position,long id){
72              if(unlockApps.get(position).packageName.equals(
73              "cn.itcast.mobliesafe")){
74                  return;
75              }
76              //给应用加锁，播放一个动画效果
77              TranslateAnimation ta=new TranslateAnimation(
78              Animation.RELATIVE_TO_SELF,0,Animation.RELATIVE_TO_SELF,1.0f,
79              Animation.RELATIVE_TO_SELF,0,Animation.RELATIVE_TO_SELF,0);
80              ta.setDuration(300);
81              view.startAnimation(ta);
82              new Thread(){
83                  public void run() {
84                      try{
85                          Thread.sleep(300);
86                      }catch(InterruptedException e){
87                          e.printStackTrace();
88                      }
```

```
89                    getActivity().runOnUiThread(new Runnable(){
90                        @Override
91                        public void run(){
92                            //程序锁信息被加入到数据库了
93                            dao.insert(unlockApps.get(position).packageName);
94                            unlockApps.remove(position);
95                            adapter.notifyDataSetChanged();//通知界面更新
96                        }
97                    });
98                };
99            }.start();
100        }
101    });
102   }
103 }
```

代码说明：

- 第 9～25 行 Handler 对象中处理 fillData()方法中发送的消息，先将集合中的数据全部清除，再将接收到的数据添加到集合中，如果没有创建 ListView 的 Adapter，则先创建并将集合传递过去，将未加锁应用信息显示在界面上，如果已经创建了 Adapter 则直接刷新界面。
- 第 27～33 行的 onCreateView()方法，用于初始化未加锁布局以及控件，Fragment 打开时先执行该方法。
- 第 36～37 行代码首先得到数据库对象和手机中安装的所有应用列表。
- 第 41～47 行代码，在创建数据库的添加和删除方法时做了一个操作，通知内容观察者某个 uri 的数据发生了改变，在这里使用该 uri 注册一个内容观察者，如果数据库内容发生了改变，会执行 onChange()方法，在该方法中重新执行 fillData()方法刷新界面。
- 第 49～66 行的 fillData()方法中遍历应用是否在数据库中，如果不在则添加到 ArrayList 集合中，设置标志为 false，代表该应用未加锁，并发送一条 Message。
- 第 67～102 行代码为 ListView 条目点击监听，首先判断点击的条目信息是否为本应用，如果是则直接返回不再执行，否则，执行一个动画效果。点击其他条目时，条目会从右向左滑动消失，并将该应用添加到数据库中（表示该应用已加锁）重新刷新界面。

4. 已加锁逻辑代码

已加锁界面和未加锁界面的逻辑代码非常相似，所做的操作基本相同，这里不再重复介绍，具体代码如【文件 9-32】所示。

【文件 9-32】AppLockFragment.java

```
1 public class AppLockFragment extends Fragment{
2     private TextView mLockTV;
3     private ListView mLockLV;
```

```
4    private AppLockDao dao;
5    List<AppInfo> mLockApps=new ArrayList<AppInfo>();
6    private AppLockAdapter adapter;
7    private Uri uri=Uri.parse("content://com.itcast.mobilesafe.applock");
8    private Handler mHandler=new Handler(){
9        public void handleMessage(android.os.Message msg){
10           switch(msg.what){
11           case 10:
12               mLockApps.clear();
13               mLockApps.addAll((List<AppInfo>)msg.obj);
14               if(adapter==null){
15                   adapter=new AppLockAdapter(mLockApps,getActivity());
16                   mLockLV.setAdapter(adapter);
17               }else{
18                   adapter.notifyDataSetChanged();
19               }
20               mLockTV.setText("加锁应用"+mLockApps.size()+"个");
21               break;
22           }
23        };
24   };
25   private List<AppInfo> appInfos;
26   @Override
27   public View onCreateView(LayoutInflater inflater,ViewGroup container,
28   Bundle savedInstanceState){
29       View view=inflater.inflate(R.layout.fragment_applock,null);
30       mLockTV=(TextView) view.findViewById(R.id.tv_lock);
31       mLockLV=(ListView) view.findViewById(R.id.lv_lock);
32       return view;
33   }
34   @Override
35   public void onResume(){
36       dao=new AppLockDao(getActivity());
37       appInfos=AppInfoParser.getAppInfos(getActivity());
38       fillData();
39       initListener();
40       getActivity().getContentResolver().registerContentObserver(uri, true,
41       new ContentObserver(new Handler()){
42           @Override
```

```java
43          public void onChange(boolean selfChange){
44              fillData();
45          }
46      });
47      super.onResume();
48  }
49  private void fillData(){
50      final List<AppInfo> aInfos=new ArrayList<AppInfo>();
51      new Thread(){
52          public void run(){
53              for (AppInfo appInfo:appInfos){
54                  if(dao.find(appInfo.packageName)){
55                      //已加锁
56                      appInfo.isLock=true;
57                      aInfos.add(appInfo);
58                  }
59              }
60              Message msg=new Message();
61              msg.obj=aInfos;
62              msg.what=10;
63              mHandler.sendMessage(msg);
64          };
65      }.start();
66  }
67  private void initListener(){
68      mLockLV.setOnItemClickListener(new OnItemClickListener(){
69          @Override
70          public void onItemClick(AdapterView<?> parent,View view,
71          final int position,long id){
72              //播放一个动画效果
73              TranslateAnimation ta=new TranslateAnimation(
74              Animation.RELATIVE_TO_SELF,0,Animation.RELATIVE_TO_SELF,-1.0f,
75              Animation.RELATIVE_TO_SELF,0,Animation.RELATIVE_TO_SELF,0);
76              ta.setDuration(300);
77              view.startAnimation(ta);
78              new Thread(){
79                  public void run(){
80                      try{
81                          Thread.sleep(300);
```

```
82                  }catch (InterruptedException e){
83                      e.printStackTrace();
84                  }
85                  getActivity().runOnUiThread(new Runnable() {
86                      @Override
87                      public void run() {
88                          //删除数据库的包名
89                          dao.delete(mLockApps.get(position).packageName);
90                          //更新界面
91                          mLockApps.remove(position);
92                          adapter.notifyDataSetChanged();
93                      }
94                  });
95              };
96          }.start();
97      }
98    });
99  }
100 }
```

如上代码所示，已加锁界面的逻辑与未加锁的逻辑完全一致，其中的差别就是在点击某个条目时将该条目信息从数据库中移除。

5. 数据适配器

未加锁界面和已加锁界面使用的数据适配器都是 AppLockAdapter，它接收一个 List 集合，不同界面调用 Adapter 时，传输的数据集合不同，例如未加锁界面调用时传递过来的集合数据是未加锁应用信息，已加锁界面调用时传递过来的集合数据是已加锁应用信息。数据适配器代码如【文件 9-33】所示。

【文件 9-33】AppLockAdapter.java

```
1  /**此类可复用，未加锁和已加锁都可以用此Adapter*/
2  public class AppLockAdapter extends BaseAdapter{
3      private List<AppInfo> appInfos;
4      private Context context;
5      /**
6       * 构造方法
7       * @param appInfos
8       * @param context
9       */
10     public AppLockAdapter(List<AppInfo> appInfos,Context context){
11         super();
```

```java
12        this.appInfos=appInfos;
13        this.context=context;
14    }
15    @Override
16    public int getCount(){
17        //TODO Auto-generated method stub
18        return appInfos.size();
19    }
20    @Override
21    public Object getItem(int position){
22        //TODO Auto-generated method stub
23        return appInfos.get(position);
24    }
25    @Override
26    public long getItemId(int position){
27        //TODO Auto-generated method stub
28        return position;
29    }
30    @Override
31    public View getView(int position,View convertView,ViewGroup parent){
32        ViewHolder holder;
33        if(convertView!=null && convertView instanceof RelativeLayout){
34            holder=(ViewHolder) convertView.getTag();
35        }else{
36            holder=new ViewHolder();
37            convertView = View.inflate(context, R.layout.item_list_applock, null);
38            holder.mAppIconImgv=(ImageView) convertView.findViewById(
39            R.id.imgv_appicon);
40            holder.mAppNameTV=(TextView) convertView.findViewById
41            (R.id.tv_appname);
42            holder.mLockIcon = (ImageView) convertView.findViewById(R.id.imgv_lock);
43            convertView.setTag(holder);
44        }
45        final AppInfo appInfo=appInfos.get(position);
46        holder.mAppIconImgv.setImageDrawable(appInfo.icon);
47        holder.mAppNameTV.setText(appInfo.appName);
48        if(appInfo.isLock){
49            //表示当前应用已经加锁
50            holder.mLockIcon.setBackgroundResource(R.drawable.applock_icon);
```

```
51        }else{
52            //当前应用未加锁
53            holder.mLockIcon.setBackgroundResource(R.drawable.appunlock_icon);
54        }
55        return convertView;
56    }
57    static class ViewHolder{
58        TextView mAppNameTV;
59        ImageView mAppIconImgv;
60        /**控制图片显示加锁还是不加锁*/
61        ImageView mLockIcon;
62    }
63 }
```

9.6.8 程序锁服务

界面逻辑已经全部完成,最后一步要实现的是程序锁功能中的核心内容——程序锁服务。需要通过该服务来获取任务栈的信息来判断当前开启的哪个应用,是否需要弹出密码锁界面。程序锁服务代码如【文件9-34】所示。

【文件9-34】AppLockService.java

```
1  /**
2   * 程序锁服务
3   * @author admin
4   */
5  public class AppLockService extends Service{
6      /** 是否开启程序锁服务的标志 */
7      private boolean flag=false;
8      private AppLockDao dao;
9      private Uri uri = Uri.parse("content://com.itcast.mobilesafe.applock");
10     private List<String> packagenames;
11     private Intent intent;
12     private ActivityManager am;
13     private List<RunningTaskInfo> taskInfos;
14     private RunningTaskInfo taskInfo;
15     private String pacagekname;
16     private String tempStopProtectPackname;
17     private AppLockReceiver receiver;
18     private MyObserver observer;
19     @Override
20     public IBinder onBind(Intent intent){
```

```
21        return null;
22    }
23    @Override
24    public void onCreate(){
25        //创建 AppLockDao 实例
26        dao=new AppLockDao(this);
27        observer=new MyObserver(new Handler());
28        getContentResolver().registerContentObserver(uri,true,observer);
29        //获取数据库中的所有包名
30        packagenames=dao.findAll();
31        receiver=new AppLockReceiver();
32        IntentFilter filter=new IntentFilter("cn.itcast.mobliesafe.applock");
33        filter.addAction(Intent.ACTION_SCREEN_ON);
34        filter.addAction(Intent.ACTION_SCREEN_OFF);
35        registerReceiver(receiver,filter);
36        //创建 Intent 实例,用来打开输入密码页面
37        intent=new Intent(AppLockService.this,EnterPswActivity.class);
38        //获取 ActivityManager 对象
39        am=(ActivityManager) getSystemService(ACTIVITY_SERVICE);
40        startApplockService();
41        super.onCreate();
42    }
43    /**
44    * 开启监控程序服务
45    */
46    private void startApplockService(){
47        new Thread(){
48            public void run(){
49                flag=true;
50                while(flag){
51                    //监视任务栈的情况,使用的打开的任务栈在集合的最前面
52                    taskInfos=am.getRunningTasks(1);
53                    //最近使用的任务栈
54                    taskInfo=taskInfos.get(0);
55                    pacagekname=taskInfo.topActivity.getPackageName();
56                    //判断这个包名是否需要被保护
57                    if(packagenames.contains(pacagekname)){
58                        //判断当前应用程序是否需要临时停止保护(输入了正确的密码)
59                        if(!pacagekname.equals(tempStopProtectPackname)){
```

```
60                    //需要保护,弹出一个输入密码的界面
61                    intent.putExtra("packagename",pacagekname);
62                    intent.setFlags(Intent.FLAG_ACTIVITY_NEW_TASK);
63                    startActivity(intent);
64                }
65            }
66            try{
67                Thread.sleep(30);
68            }catch(InterruptedException e){
69                e.printStackTrace();
70            }
71        }
72    };
73    }.start();
74 }
75 //广播接收者
76 class AppLockReceiver extends BroadcastReceiver{
77    @Override
78    public void onReceive(Context context, Intent intent){
79       if("cn.itcast.mobliesafe.applock".equals(intent.getAction())){
80          tempStopProtectPackname = intent.getStringExtra("packagename");
81       }else if(Intent.ACTION_SCREEN_OFF.equals(intent.getAction())){
82          tempStopProtectPackname=null;
83          //停止监控程序
84          flag=false;
85       }else if(Intent.ACTION_SCREEN_ON.equals(intent.getAction())) {
86          //开启监控程序
87          if(flag==false){
88             startApplockService();
89          }
90       }
91    }
92 }
93 //内容观察者
94 class MyObserver extends ContentObserver{
95    public MyObserver(Handler handler){
96       super(handler);
97    }
98    @Override
```

```
99      public void onChange(boolean selfChange){
100         packagenames=dao.findAll();
101         super.onChange(selfChange);
102     }
103 }
104 @Override
105 public void onDestroy(){
106     flag=false;
107     unregisterReceiver(receiver);
108     receiver=null;
109     getContentResolver().unregisterContentObserver(observer);
110     observer=null;
111     super.onDestroy();
112 }
113 }
```

代码说明：

- 第 30~35 行代码获取数据库中保存的所有包名信息。
- 第 46~74 行 startApplockService()方法中开启一个子线程实时监控被打开的程序。
- 第 52~55 行代码获取正在运行的任务栈列表并得到最近使用的应用程序的信息。注意，由于服务一直在后台运行，开启死循环一直监听任务栈操作是非常消耗内存的，因此在第 66~70 行代码让线程间隔 30 毫秒执行一次。
- 第 57~65 行代码将通过监听任务栈获得的包名与数据库中存储的包名进行对比，如果数据库中存在，则表示该应用已加锁，开启密码锁界面并将当前打开的应用包名传递过去。注意，当打开已加锁应用在密码锁界面输入密码之后，由于程序锁服务还在监听任务栈信息，会再次弹出密码锁界面。为了防止这种情况发生，需要将输入正确密码的应用进行临时取消保护。当在密码锁界面输入完成后会发送一条自定义广播传输一个 action 为 cn.itcast.mobliesafe.applock，在 76~92 行代码自定义广播接收者，当广播接收到这个 action 时获取已解锁应用的包名，并在第 57 行代码进行判断，如果当前应用包名不是已解锁应用时才打开密码锁界面。这样就防止密码锁界面重复打开。
- 第 76~92 行代码在自定义广播中判断屏幕是否锁屏或解锁，当锁屏时停止监控任务栈信息，解锁时再重新进行监控，这样会节省手机电量，使应用消耗减少。
- 第 94~103 行代码定义了一个内容观察者 MyObserver，因为数据库中的信息是在服务开启 onCreate()时得到的，如果服务运行过程中将某个应用加锁或者解锁，数据库发生变化，这时得到数据库信息不能及时更新。因此在数据库的添加和删除方法中做了一个操作：通知内容观察者某一个 uri 的内容发生变化，广播的 uri 是自定义的，"content://com.itcast.mobilesafe.applock"（在 9.6.2 小节中介绍过）。当数据库发生变化时重新从数据库中拿一遍数据，保证数据是最新的。

接下来在 AndroidManifest.xml 文件中注册程序锁服务，具体代码如下：

```
<!-- 程序锁 -->
```

```
<service
    android:name="cn.itcast.mobliesafe.chapter09.service.AppLockService"
    android:persistent="true" >
</service>
```

9.7 密 码 锁

9.7.1 密码锁界面

程序锁服务开启后，当打开手机上的某一款应用时，程序锁服务会判断该应用是否在已加锁数据库中，如果在则先打开一个输入密码的界面，只有密码输入正确才能打开该应用。密码锁界面如图 9-14 所示。

图 9-14 所示密码锁界面的布局文件如【文件 9-35】所示。

图 9-14 密码锁界面

【文件 9-35】activity_enterpsw.xml

```
<?xml version="1.0" encoding="utf-8"?>
<LinearLayout xmlns:android="http://schemas.android.
    com/apk/res/android"
    android:layout_width="match_parent"
    android:layout_height="match_parent"
    android:gravity="center"
    android:orientation="vertical">
    <ImageView
        android:id="@+id/imgv_appicon_enterpsw"
        android:layout_width="80dp"
        android:layout_height="80dp"
        android:layout_gravity="center_horizontal"/>
    <TextView
        android:id="@+id/tv_appname_enterpsw"
        style="@style/wrapcontent"
        android:layout_gravity="center_horizontal"
        android:layout_margin="10dp"
        android:textColor="@color/dark_gray"
        android:textScaleX="1.2"
        android:textSize="20sp"/>
    <LinearLayout
        android:id="@+id/ll_enterpsw"
        android:layout_width="match_parent"
        android:layout_height="45dp"
```

```xml
            android:layout_margin="10dp"
            android:background="@drawable/coner_white_rec"
            android:orientation="horizontal" >
            <ImageView
                android:layout_width="wrap_content"
                android:layout_height="match_parent"
                android:layout_weight="0.5"
                android:background="@drawable/enterpsw_icon" />
            <EditText
                android:id="@+id/et_psw_enterpsw"
                android:layout_width="match_parent"
                android:layout_height="match_parent"
                android:layout_weight="3"
                android:background="@null"
                android:hint="请输入防盗密码"
                android:inputType="textPassword" />
            <ImageView
                android:id="@+id/imgv_go_enterpsw"
                android:layout_width="wrap_content"
                android:layout_height="match_parent"
                android:layout_weight="0.5"
                android:background="@drawable/go_greenicon" />
        </LinearLayout>
</LinearLayout>
```

如上述代码所示，该界面主要有五个控件，输入框的上方有一个 ImageView 和一个 TextView 分别用于显示应用的图标和名称。下方是一个线性布局，该布局中包含两个 ImageView 和一个 EditText，第一个 ImageView 用于显示输入框左侧图片，第二个 ImageView 用于显示右侧的图片，EditText 用于输入防盗密码。

9.7.2 密码锁逻辑

该界面的主要作用为当用户打开已经被加锁的应用时，会先弹出密码锁界面，然后在该界面输入密码，如果密码正确则进入应用，如果密码不正确则停留在该界面，密码锁界面逻辑具体代码如【文件 9-36】所示。

【文件 9-36】EnterPswActivity.java

```
1  public class EnterPswActivity extends Activity implements OnClickListener{
2      private ImageView mAppIcon;
3      private TextView mAppNameTV;
4      private EditText mPswET;
5      private ImageView mGoImgv;
```

```java
6    private LinearLayout mEnterPswLL;
7    private SharedPreferences sp;
8    private String password;
9    private String packagename;
10   @Override
11   protected void onCreate(Bundle savedInstanceState){
12       super.onCreate(savedInstanceState);
13       requestWindowFeature(Window.FEATURE_NO_TITLE);
14       setContentView(R.layout.activity_enterpsw);
15       sp=getSharedPreferences("config",MODE_PRIVATE);
16       password=sp.getString("PhoneAntiTheftPWD",null);
17       Intent intent=getIntent();
18       packagename=intent.getStringExtra("packagename");
19       PackageManager pm=getPackageManager();
20       initView();
21       try{
22           mAppIcon.setImageDrawable(pm.getApplicationInfo(packagename,0).
23               loadIcon(pm));
24           mAppNameTV.setText(pm.getApplicationInfo(packagename,0).
25               loadLabel(pm).toString());
26       }catch(NameNotFoundException e){
27           e.printStackTrace();
28       }
29   }
30   /**
31    *  初始化控件
32    */
33   private void initView(){
34       mAppIcon=(ImageView) findViewById(R.id.imgv_appicon_enterpsw);
35       mAppNameTV=(TextView) findViewById(R.id.tv_appname_enterpsw);
36       mPswET=(EditText) findViewById(R.id.et_psw_enterpsw);
37       mGoImgv=(ImageView) findViewById(R.id.imgv_go_enterpsw);
38       mEnterPswLL=(LinearLayout) findViewById(R.id.ll_enterpsw);
39       mGoImgv.setOnClickListener(this);
40   }
41   @Override
42   public void onClick(View v){
43       switch (v.getId()){
44       case R.id.imgv_go_enterpsw:
```

```
45          //比较密码
46          String inputpsw=mPswET.getText().toString().trim();
47          if(TextUtils.isEmpty(inputpsw)){
48              startAnim();
49              Toast.makeText(this,"请输入密码!",0).show();
50              return;
51          }else{
52              if(!TextUtils.isEmpty(password)){
53                  if(MD5Utils.encode(inputpsw).equals(password)){
54                      //发送自定义的广播消息。
55                      Intent intent=new Intent();
56                      intent.setAction("cn.itcast.mobliesafe.applock");
57                      intent.putExtra("packagename",packagename);
58                      sendBroadcast(intent);
59                      finish();
60                  }else{
61                      startAnim();
62                      Toast.makeText(this,"密码不正确!",0).show();
63                      return;
64                  }
65              }
66          }
67          break;
68      }
69  }
70  private void startAnim(){
71      Animation animation = AnimationUtils.loadAnimation(this, R.anim.shake);
72      mEnterPswLL.startAnimation(animation);
73  }
74 }
```

代码说明：

- 第 11~29 行的 onCreate()方法用于初始化界面，其中第 15~16 行代码使用 SharedPreferences 取出的 String 类型的数据就是在第一个模块中设置的密码；第 17~18 行代码得到由 AppLockService 类传递过来的 packagename；第 19~28 行代码通过 packagename 得到该应用的图标和应用名称。

- 第 44~67 行代码为"确定"按钮点击监听，首先判断输入框是否为空，如果为空则进行提示，如不为空则判断之前是否设置过密码，如果设置过密码并且当前输入密码与之前设置的密码一致，则发送一条自定义广播，自定义广播的接收者在 AppLockService 类中已经介绍过；如果当前输入的密码和之前设置的密码不一致则运行动画，使输入框左右晃动，并提示密码不正确。

- 第 70～73 行代码定义输入密码框中使用的晃动动画,该动画在 res 目录下的 shake.xml 文件中设置的,具体代码如【文件 9-37】所示。

【文件 9-37】res/anim/shake.xml

```xml
<?xml version="1.0" encoding="utf-8"?>
<translate xmlns:android="http://schemas.android.com/apk/res/android"
    android:duration="1000"
    android:fromXDelta="0"
    android:interpolator="@anim/cycle_7"
    android:toXDelta="10"/>
```

当密码锁界面完成之后可以进行测试。由于本项目中服务的开启与关闭都在设置中心中,这里还没有讲解,编程者可以在程序中加入测试代码先开启服务。进入"高级工具"模块的程序锁功能,在未加锁页面中点击一款应用进行加锁,会看到被点击的条目滑动消失并且在已加锁界面中显示出来。最后在桌面上打开该应用会先弹出输入密码的界面,输入正确的密码之后才能成功打开应用。这时说明程序锁功能已经完成了,它可以有效地保护我们的隐私不被别人随意查看。

本 章 小 结

本章主要是针对高级工具模块进行讲解,首先针对该模块所有功能、代码结构进行介绍,然后分步讲解该模块的四大功能:号码归属地查询、短信备份、短信还原与程序锁。该模块的代码量较大,实现的功能较多,也比较实用,编程者需要动手将这些功能全部实现,加强代码编写能力,如遇到难以理解的逻辑或功能,可以先将程序打断点观察程序执行逻辑。

【面试精选】
1. 请简述一下 ViewPager 的缓存机制。
2. 请问线程间通信有几种方式?
扫描右方二维码,查看面试题答案!

第10章 设置中心模块

学习目标
- 了解设置中心模块功能；
- 掌握如何自定义组合控件。

在 Android 系统中，大部分软件都有一个"设置"菜单，该菜单通常用于对某个程序的个别功能进行设置，如设置皮肤、设置短信通知等。同样，在手机安全卫士中也有一个设置模块，该模块用于设置黑名单拦截及程序锁是否开启。本章将针对设置中心模块进行详细讲解。

10.1 模块概述

10.1.1 功能介绍

设置中心模块用于设置黑名单拦截及程序锁设置是否开启，默认情况下黑名单设置是开启的，程序锁设置是关闭的，当点击右侧的 ToggleButton 按钮可以进行状态的切换，该功能界面效果如图 10-1 所示。

图 10-1　软件管家界面

10.1.2 代码结构

设置中心模块是手机安全卫士中最简单的一个，该模块中主要使用了三个类，具体如图 10-2 所示。

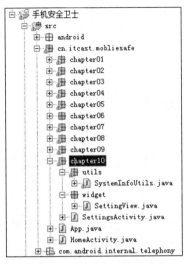

图 10-2 chapter10 代码结构

下面按照结构顺序依次介绍 chapter10 包中的文件，具体如下：

- SystemInfoUtils.java：该类用于判断程序锁服务（AppLockService）是否运行。
- SettingView.java：自定义控件类，用于定义设置中心中的条目。
- SettingsActivity.java：设置中心界面的逻辑代码。

10.2 设置中心

10.2.1 自定义控件

1. 自定义控件布局

通过前面的功能介绍可知，设置中心界面有两个小条目，每个条目有两个 TextView、一个 ToggleButton 以及一个 View 对象。由于通过代码控制自定义控件的位置非常麻烦，因此，可以先定义一个 Item 布局作为自定义控件的布局，如图 10-3 所示。

图 10-3 所示自定义控件对应的布局文件如【文件 10-1】所示。

【文件 10-1】ui_settings_view.xml

图 10-3 自定义控件布局

```xml
<?xml version="1.0" encoding="utf-8"?>
<RelativeLayout xmlns:android="http://schemas.android.com/apk/res/android"
    android:layout_width="match_parent"
    android:layout_height="55dp"
    android:orientation="vertical">
    <TextView
        android:id="@+id/tv_setting_title"
        style="@style/textview16sp"
        android:padding="3dp"
```

```xml
        android:text="黑名单设置"
        android:textColor="@color/bright_blue" />
<TextView
        android:id="@+id/tv_setting_status"
        style="@style/textview14sp"
        android:layout_below="@+id/tv_setting_title"
        android:padding="3dp"
        android:text="黑名单拦截已开启"
        android:textColor="@color/dark_gray" />
<ToggleButton
        android:id="@+id/toggle_setting_status"
        android:layout_width="70dp"
        android:layout_height="30dp"
        android:layout_alignParentRight="true"
        android:layout_centerVertical="true"
        android:layout_marginRight="27dp"
        android:background="@drawable/toggle_btn_blue_selector"
        android:textOff=""
        android:textOn=""/>
<View
        android:layout_width="match_parent"
        android:layout_height="1.5px"
        android:background="@color/black30"
        android:layout_alignParentBottom="true"/>
</RelativeLayout>
```

上述布局中,ToggleButton 按钮使用了一个背景选择器,当按钮处于选中状态时,为按钮指定一张蓝色的背景图片(toggle_btn_blue_p.png,开启状态),当按钮未被选中时,为按钮指定一张灰色的背景图(swtich_btn_off.png,关闭状态)。背景选择器的代码如【文件 10-2】所示。

【文件 10-2】 res/drawable/toggle_btn_blue_selector.xml

```xml
<?xml version="1.0" encoding="utf-8"?>
<selector xmlns:android="http://schemas.android.com/apk/res/android" >
    <item android:state_checked="true" android:drawable="@drawable/toggle_btn_blue_p"/>
    <item android:state_checked="false" android:drawable="@drawable/switch_btn_off"/>
</selector>
```

2. 自定义控件属性

由于自定义控件中有设置标题、状态开启或关闭、Toggle 是否被选中等信息,因此需要在 attrs.xml 中定义对应的四个属性，具体代码如【文件 10-3】所示。

【文件 10-3】 res/values/attrs.xml

```xml
<?xml version="1.0" encoding="utf-8"?>
<resources>
    <declare-styleable name="SettingView">
        <attr name="settitle" format="string"/>
        <attr name="status_on" format="string"/>
        <attr name="status_off" format="string"/>
        <attr name="status_ischecked" format="boolean"/>
    </declare-styleable>
</resources>
```

上述代码中，<declare-styleable>节点表示自定义属性，<attr>节点中的内容就是具体的属性，name 表示属性名称，format 表示属性的数据类型。

3. 实现自定义控件

自定义控件的布局以及所需的属性都已定义完成，接下来通过代码实现自定义控件，在定义控件时，需要继承 RelativeLayout，并创建自定义控件类的构造方法，具体代码如【文件 10-4】所示。

【文件 10-4】 SettingView.java

```
1  public class SettingView extends RelativeLayout{
2      private String setTitle="";
3      private String status_on="";
4      private String status_off="";
5      private TextView mSettingTitleTV;
6      private TextView mSettingStatusTV;
7      private ToggleButton mToggleBtn;
8      private boolean isChecked;
9      private OnCheckedStatusIsChanged onCheckedStatusIsChanged;
10     public SettingView(Context context){
11         super(context);
12         init(context);
13     }
14     public SettingView(Context context,AttributeSet attrs, int defStyle){
15         super(context,attrs,defStyle);
16         init(context);
17     }
18     public SettingView(Context context,AttributeSet attrs){
19         super(context,attrs);
20         //拿到属性对象的值
```

```java
21      TypedArray mTypedArray=context.obtainStyledAttributes(attrs,
22      R.styleable.SettingView);
23      setTitle=mTypedArray.getString(R.styleable.SettingView_settitle);
24      status_on=mTypedArray.getString(R.styleable.SettingView_status_on);
25      status_off=mTypedArray.getString(R.styleable.SettingView_status_off);
26      isChecked=mTypedArray.getBoolean(R.styleable.SettingView_status_ischecked,
27      false);
28      mTypedArray.recycle();
29      init(context);
30      setStatus(status_on,status_off,isChecked);
31  }
32  /**
33   * 初始化控件
34   * @param context
35   */
36  private void init(Context context){
37      View view= View.inflate(context,R.layout.ui_settings_view,null);
38      this.addView(view);
39      mSettingTitleTV=(TextView) findViewById(R.id.tv_setting_title);
40      mSettingStatusTV=(TextView) findViewById(R.id.tv_setting_status);
41      mToggleBtn=(ToggleButton) findViewById(R.id.toggle_setting_status);
42      mSettingTitleTV.setText(setTitle);
43      mToggleBtn.setOnCheckedChangeListener(new OnCheckedChangeListener(){
44          @Override
45          public void onCheckedChanged(CompoundButton buttonView,boolean isChecked){
46              setChecked(isChecked);
47              onCheckedStatusIsChanged.onCheckedChanged(SettingView.this,isChecked);
48          }
49      });
50  }
51  /**
52   * 返回组合控件是否选中
53   * @return
54   */
55  public boolean isChecked(){
56      return mToggleBtn.isChecked();
57  }
58  /**
59   * 设置组合控件的选中方式
```

```java
60      * @param checked
61      */
62     public void setChecked(boolean checked){
63        mToggleBtn.setChecked(checked);
64        isChecked=checked;
65        if(checked){
66           if(!TextUtils.isEmpty(status_on)){
67              mSettingStatusTV.setText(status_on);
68           }
69        }else{
70           if(!TextUtils.isEmpty(status_off)){
71              mSettingStatusTV.setText(status_off);
72           }
73        }
74     }
75     /**
76      * 设置自定义控件的描述
77      * @param text
78      */
79     public void setStatus(String status_on,String status_off,boolean checked){
80        if(checked){
81           mSettingStatusTV.setText(status_on);
82        }else{
83           mSettingStatusTV.setText(status_off);
84        }
85        mToggleBtn.setChecked(checked);
86     }
87     /**
88      * 设置状态变化监听
89      */
90     public void setOnCheckedStatusIsChanged(OnCheckedStatusIsChanged
91     onCheckedStatusIsChanged){
92        this.onCheckedStatusIsChanged = onCheckedStatusIsChanged;
93     }
94     /**
95      * 回调接口
96      * @author admin
97      */
98     public interface OnCheckedStatusIsChanged{
```

```
99          /**
100          * 当前 View 的 check 状态发生变化时会调用此方法
101          * @param view
102          * @param isChecked
103          */
104         void onCheckedChanged(View view,boolean isChecked);
105     }
106 }
```

代码说明：

- 第 18～31 行的 SettingView()方法用于获取属性，首先通过 context.obtainStyledAttributes (attrs,R.styleable.SettingView)创建 TypedArray 对象，然后通过该对象获取 attrs.xml 文件中定义的属性值，最后回收 TypedArray 对象。
- 第 36～50 行的 init()方法用于对 ui_settings_view.xml 文件中的控件进行初始化，并为 ToggleButton 按钮设置切换状态的监听事件。第 47 行代码调用 onCheckedStatusIsChanged 回调接口中的 onCheckedChanged()方法。
- 第 62～74 行的 setChecked()方法用于实现组合控件的选中方式，黑名单拦截是否开启与程序锁是否开启都可以调用该方法。
- 第 79～86 行的 setStatus()方法用于设置组合控件(黑名单拦截或程序锁)是否选中，如果是选中状态，则 mSettingStatusTV 显示文本为已开启（status_on），否则显示已关闭（status_off）。
- 第 90～93 行的 setOnCheckedStatusIsChanged()方法用于设置状态监听，并传递了一个接口对象 onCheckedStatusIsChanged。
- 第 98～105 行代码定义了一个状态改变的回调接口 OnCheckedStatusIsChanged。

10.2.2 设置中心界面

设置中心 UI 在布局文件中引入两个自定义组合控件 SettingView，当使用自定义控件时，首先需要为自定义控件声明命名空间。在 Android 系统中，声明控件的命名空间格式为"xmlns:前缀=http://schemas.android.com/apk/res/控件所在的包路径"，其中 xmlns 表示命名空间（xml namespace），冒号后面是这个引用的别名，又称前缀，可以自己定义。

当引用自定义控件时，需要使用控件的全路径，例如在布局文件中引入 SettingView 控件时，需要使用"<cn.itcast.mobliesafe.chapter10.widget.SettingView>"，具体代码如【文件 10-5】所示。

【文件 10-5】activity_settings.xml

```
<?xml version="1.0" encoding="utf-8"?>
<LinearLayout xmlns:android="http://schemas.android.com/apk/res/android"
    xmlns:itcast="http://schemas.android.com/apk/res/cn.itcast.mobliesafe"
    android:layout_width="match_parent"
    android:layout_height="match_parent"
    android:orientation="vertical">
```

```xml
<include layout="@layout/titlebar"/>
<cn.itcast.mobliesafe.chapter10.widget.SettingView
    android:id="@+id/sv_blacknumber_set"
    android:layout_width="match_parent"
    android:layout_height="55dp"
    itcast:status_on="黑名单拦截已开启"
    itcast:status_off="黑名单拦截已关闭"
    itcast:settitle="黑名单拦截设置"/>
<cn.itcast.mobliesafe.chapter10.widget.SettingView
    android:id="@+id/sv_applock_set"
    android:layout_width="match_parent"
    android:layout_height="55dp"
    itcast:status_on="程序锁已开启"
    itcast:status_off="程序锁已关闭"
    itcast:settitle="程序锁设置"/>
</LinearLayout>
```

上述布局文件中,引入了一个标题栏布局(titlebar.xml),以及两个自定义控件(SettingView)。自定义控件与普通控件一样,也可以设置 id、layout_width、layout_height 等属性。

10.2.3 工具类

程序锁是否开启是通过程序锁服务(AppLockService)控制的,因此可以专门定义一个工具类用于判断程序锁服务是否运行,具体代码如【文件 10-6】所示。

【文件 10-6】SystemInfoUtils.java

```java
1  public class SystemInfoUtils{
2    /**
3     * 判断一个服务是否处于运行状态
4     * @param context 上下文
5     * @return
6     */
7    public static boolean isServiceRunning(Context context,String className){
8      ActivityManager am=(ActivityManager) context.getSystemService(
9      Context.ACTIVITY_SERVICE);
10     List<RunningServiceInfo> infos=am.getRunningServices(200);
11     for(RunningServiceInfo info:infos){
12       String serviceClassName=info.service.getClassName();
13       if(className.equals(serviceClassName)){
14         return true;
15       }
16     }
```

```
17        return false;
18    }
19 }
```

上述代码定义了一个 isServiceRunning(Context context,String className)方法，该方法用于判断某个服务是否处于运行状态，它接收两个参数，分别为 context 和 className（类名）。在进行判断的过程中，首先需要获取 ActivityManager 对象，然后通过该对象的 getRunningServices()方法获取当前正在运行的所有服务，最后通过 for 循环遍历每个服务并通过名称进行比对，如果相同则返回 true，否则返回 false。

10.2.4 设置中心逻辑

设置中心的 UI 以及判断服务是否运行的工具类已经编写完成，接下来在界面逻辑代码（SettingsActivity）中进行调用，完成黑名单以及程序锁的设置，具体代码如【文件 10-7】所示。

【文件 10-7】SettingsActivity.java

```
1  public class SettingsActivity extends Activity implements OnClickListener,
2  OnCheckedStatusIsChanged {
3      private SettingView mBlackNumSV;
4      private SettingView mAppLockSV;
5      private SharedPreferences mSP;
6      private boolean running;
7      private Intent intent;
8      @Override
9      protected void onCreate(Bundle savedInstanceState){
10         super.onCreate(savedInstanceState);
11         requestWindowFeature(Window.FEATURE_NO_TITLE);
12         setContentView(R.layout.activity_settings);
13         mSP=getSharedPreferences("config",MODE_PRIVATE);
14         initView();
15     }
16     private void initView(){
17         findViewById(R.id.rl_titlebar).setBackgroundColor(
18                 getResources().getColor(R.color.bright_blue));
19         ImageView mLeftImgv=(ImageView) findViewById(R.id.imgv_leftbtn);
20         ((TextView) findViewById(R.id.tv_title)).setText("设置中心");
21         mLeftImgv.setOnClickListener(this);
22         mLeftImgv.setImageResource(R.drawable.back);
23         mBlackNumSV=(SettingView) findViewById(R.id.sv_blacknumber_set);
24         mAppLockSV=(SettingView) findViewById(R.id.sv_applock_set);
25         mBlackNumSV.setOnCheckedStatusIsChanged(this);
```

```java
26       mAppLockSV.setOnCheckedStatusIsChanged(this);
27    }
28    @Override
29    protected void onStart(){
30       running=SystemInfoUtils.isServiceRunning(this,
31       "cn.itcast.mobliesafe.chapter09.service.AppLockService");
32       mAppLockSV.setChecked(running);
33       mBlackNumSV.setChecked(mSP.getBoolean("BlackNumStatus",true));
34       super.onStart();
35    }
36    @Override
37    public void onClick(View v){
38       switch (v.getId()){
39       case R.id.imgv_leftbtn:
40          finish();
41          break;
42       }
43    }
44    @Override
45    public void onCheckedChanged(View view,boolean isChecked){
46       switch (view.getId()){
47       case R.id.sv_blacknumber_set:
48          saveStatus("BlackNumStatus",isChecked);
49          break;
50       case R.id.sv_applock_set:
51          saveStatus("AppLockStatus",isChecked);
52          //开启或者关闭程序锁
53          if(isChecked){
54             intent=new Intent(this,AppLockService.class);
55             startService(intent);
56          }else{
57             stopService(intent);
58          }
59          break;
60       }
61    }
62    private void saveStatus(String keyname,boolean isChecked){
63       if(!TextUtils.isEmpty(keyname)){
64          Editor edit = mSP.edit();
```

```
65        edit.putBoolean(keyname, isChecked);
66        edit.commit();
67    }
68 }
69}
```

代码说明：

- 第 16~27 行的 initView()方法用于初始化控件，并为黑名单按钮和程序锁按钮设置状态变化的监听器。
- 第 29~35 行代码重写了 Activity 的 onStart()方法，当 Activity 启动时会自动调用该方法，判断程序锁服务是否运行、程序锁是否开启，并将返回的结果设置给 mAppLockSV 控件和 mBlackNumSV 控件。
- 第 45~61 行的 onCheckedChanged()方法，用于监听 ToggleButton 按钮的状态改变。当点击"黑名单开启"按钮时，调用 saveStatus()方法将 BlackNumStatus 状态保存起来；当点击"程序锁开启"按钮时，调用 saveStatus()方法将 AppLockStatus 状态保存起来，并通过 if 语句判断程序锁服务是开启状态还是关闭状态，如果是关闭状态则开启服务，否则关闭服务。
- 第 62~68 行的 saveStatus()方法用于保存黑名单拦截设置以及程序锁设置是否开启。

本 章 小 结

本章主要是针对设置中心模块进行讲解，首先针对该模块的功能、代码结构进行介绍，然后讲解了设置中心的自定义组合控件，最后实现设置中心的逻辑代码。该模块功能比较简单，只需要按照步骤完成模块开发即可。

【面试精选】
1. 请问 Android 中如何自定义控件？
2. 请问如何监听 EditText 控件中的数据变化？
扫描右方二维码，查看面试题答案！